Vector Analysis for Beginners

Vector Analysis for Beginners

Rohit Gupta

Oxford Book Company
Jaipur ▪ New Delhi

Published By

Oxford Book Company

Regd. Off. "Bony Residency", Gate No. 2
Opp. Tilak Public School, Vishweswariya Nagar,
Gopalpura Road, Jaipur - 302018, India
Ph: 0141 - 2761280, Telefax: 2761381
email - oxfordbookjpr@gmail.com
website - oxfordbookcompany.com

Delhi Office: 102, 1st Floor, Satyam House,
4327/3, Ansari Road, Darya Ganj,
New Delhi - 110002, India
Ph: 011-45652440

Distributed By

KEYA INTERNATIONAL PRESS LLP

E-mail: keyainternationalpressllp@gmail.com

First published in India, 2026

ISBN : 978-93-5524-700-1

Typeset by Shivangi Computers, Jaipur

Printed and bound in India by Thomson Press Ltd.,
New Delhi

Contents

Preface

Mathematics serves as the foundation of countless scientific and engineering disciplines, and among its many branches, vector analysis holds a distinctive and indispensable place. The ability to analyze and manipulate vectors is crucial in physics, engineering, computer science, and applied mathematics. Vector Analysis for Beginners is designed to introduce readers to the fundamental concepts of vector algebra and calculus in a systematic and accessible manner.

This book is tailored for students, educators, and professionals who seek a clear and concise understanding of vector analysis without the burden of excessive theoretical complexity. It provides a structured approach that balances theoretical explanations with practical applications, ensuring a well-rounded comprehension of the subject. Starting with the basics of vectors, their properties, and operations, the book gradually builds towards more advanced topics such as gradient, divergence, curl, and line, surface, and volume integrals.

One of the key strengths of this book lies in its intuitive presentation of vector concepts. Complex ideas are broken down into simple, digestible explanations, supported by illustrative diagrams, step-by-step derivations, and numerous solved examples. Each chapter is designed to reinforce learning by including exercises that encourage readers to apply their knowledge and develop problem-solving skills.

In addition to covering classical vector algebra and calculus, this book highlights the practical applications of vector analysis in physics, fluid mechanics, electromagnetism, and engineering. By integrating real-world examples, readers will gain a deeper appreciation of the relevance of vectors in diverse scientific and technological fields.

Furthermore, this book aims to cultivate an intuitive understanding of vector spaces and transformations, enabling readers to confidently navigate more advanced mathematical topics in the future. The logical progression of concepts ensures that even those with minimal mathematical background can grasp and apply the principles of vector analysis with confidence.

Vector Analysis for Beginners is more than just an introductory textbook—it is a gateway to a deeper exploration of mathematical concepts that are fundamental to modern science and engineering. Whether you are a student embarking on your mathematical journey, an educator seeking a comprehensive teaching resource, or a professional looking to refresh your knowledge, this book will serve as an essential guide, making the study of vectors both engaging and accessible.

— ***Rohit Gupta***

Chapter 1

Paraboloid

In mathematics, a paraboloid is a quadric surface of special kind. There are two kinds of paraboloids: elliptic and hyperbolic.

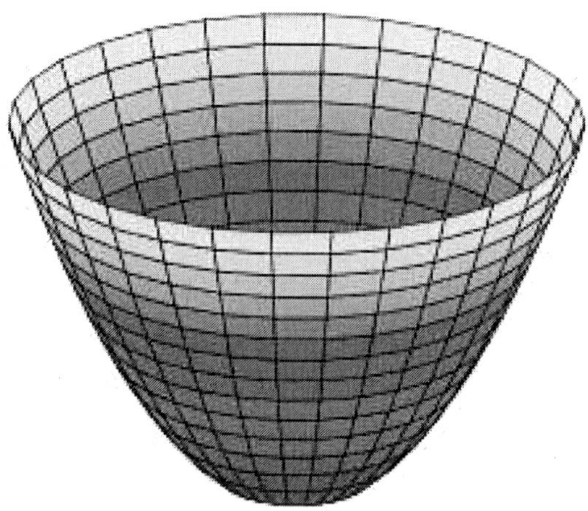

Figure: *Paraboloid of revolution*

The elliptic paraboloid is shaped like an oval cup and can have a maximum or minimum point. In a suitable coordinate system with three axes x, y, and z, it can be represented by the equation

$$\frac{z}{c} = \frac{x^2}{a^2} + \frac{y^2}{b^2}.$$

where a and $b$are constants that dictate the level of curvature in the x-z- and y-z planes respectively. This is an elliptic paraboloid which opens upward.

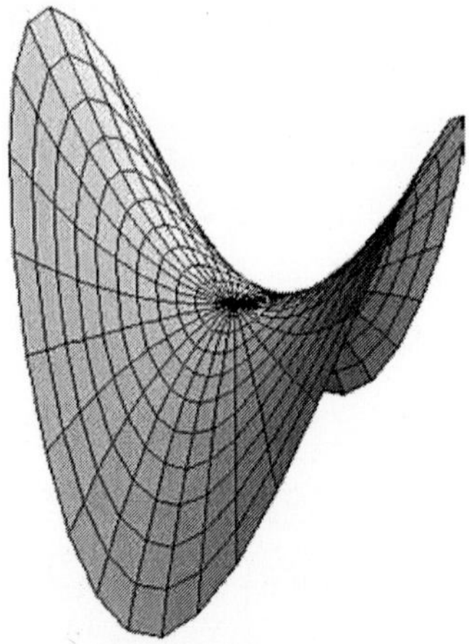

Figure: *Hyperbolic paraboloid*

The hyperbolic paraboloid (not to be confused with a hyperboloid) is a doubly ruled surface shaped like a saddle. In a suitable coordinate system, a hyperbolic paraboloid can be represented by the equation

$$\frac{z}{c} = \frac{y^2}{b^2} - \frac{x^2}{a^2}.$$

For c>0, this is a hyperbolic paraboloid that opens up along the x-axis and down along the y-axis (i.e., the parabola in the plane x=0 opens upward and the parabola in the plane y=0 opens downward).

Properties

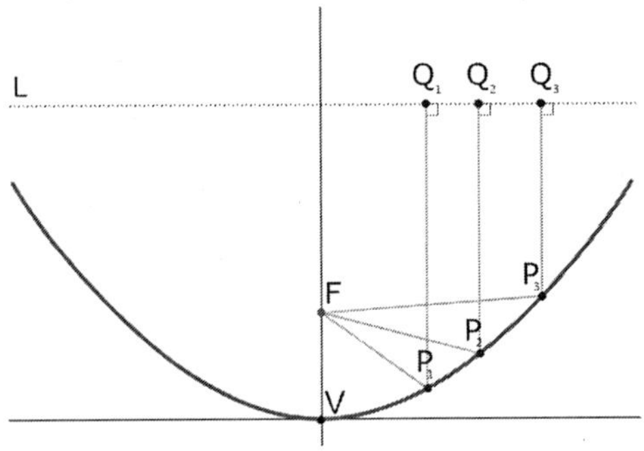

Figure: *Parallel rays coming into a circular paraboloidal mirror are reflected to the focal point, F, or vice versa.*

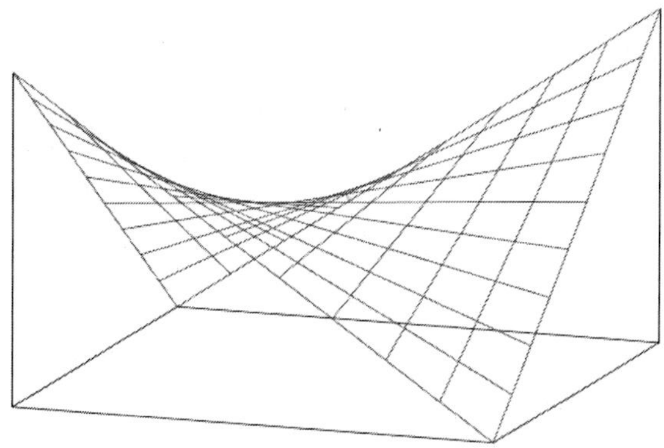

Figure: *The hyperbolic paraboloid is a doubly ruled surface, and thus can be used to construct a saddle roof from straight beams.*

With $a = b$ an elliptic paraboloid is a paraboloid of revolution: a surface obtained by revolving a parabola around its axis. It is the shape of the parabolic reflectors used in mirrors, antenna dishes, and the like; and is also the shape of the surface of a rotating liquid, a principle used in liquid mirror telescopes and in making solid telescope mirrors. This shape is also called a circular paraboloid.

There is a point called the focus (or focal point) on the axis of a circular paraboloid such that, if the paraboloid is a mirror, light from a point source at the focus is reflected into a parallel beam, parallel to the axis of the paraboloid. This also works the other way around: a parallel beam of light incident on the paraboloid parallel to its axis is concentrated at the focal point. This applies also for other waves, hence parabolic antennas.

The hyperbolic paraboloid is a doubly ruled surface: it contains two families of mutually skew lines. The lines in each family are parallel to a common plane, but not to each other.

Curvature

The elliptic paraboloid, parametrized simply as

$$\vec{\sigma}(u,v) = \left(u, v, \frac{u^2}{a^2} + \frac{v^2}{b^2} \right)$$

has Gaussian curvature

$$K(u,v)=\frac{4}{a^2b^2\left(1+\frac{4u^2}{a^4}+\frac{4v^2}{b^4}\right)^2}$$

and mean curvature

$$H(u,v)=\frac{a^2+b^2+\frac{4u^2}{a^2}+\frac{4v^2}{b^2}}{a^2b^2\left(1+\frac{4u^2}{a^4}+\frac{4v^2}{b^4}\right)^{3/2}}$$

which are both always positive, have their maximum at the origin, become smaller as a point on the surface moves further away from the origin, and tend asymptotically to zero as the said point moves infinitely away from the origin.

The hyperbolic paraboloid, when parametrized as

$$\vec{\sigma}(u,v)=\left(u,v,\frac{u^2}{a^2}-\frac{v^2}{b^2}\right)$$

has Gaussian curvature

$$K(u,v)=\frac{-4}{a^2b^2\left(1+\frac{4u^2}{a^4}+\frac{4v^2}{b^4}\right)^2}$$

and mean curvature

$$H(u,v)=\frac{-a^2+b^2-\frac{4u^2}{a^2}+\frac{4v^2}{b^2}}{a^2b^2\left(1+\frac{4u^2}{a^4}+\frac{4v^2}{b^4}\right)^{3/2}}.$$

Multiplication Table

If the hyperbolic paraboloid

$$z=\frac{x^2}{a^2}-\frac{y^2}{b^2}$$

is rotated by an angle of $\pi/4$ in the +z direction (according to the right hand rule), the result is the surface

$$z = \frac{1}{2}(x^2 + y^2)\left(\frac{1}{a^2} - \frac{1}{b^2}\right) + xy\left(\frac{1}{a^2} + \frac{1}{b^2}\right)$$

and if $a = b$ then this simplifies to

$$z = \frac{2}{a^2}xy.$$

Finally, letting $a = \sqrt{2}$, we see that the hyperbolic paraboloid

$$z = \frac{x^2 - y^2}{2}.$$

is congruent to the surface

$$z = xy$$

which can be thought of as the geometric representation (a three-dimensional nomograph, as it were) of a multiplication table.

The two paraboloidal $\mathbb{R}^2 \to \mathbb{R}$ functions

$$z_1(x, y) = \frac{x^2 - y^2}{2}$$

and

$$z_2(x, y) = xy$$

are harmonic conjugates, and together form the analytic function

$$f(z) = \frac{1}{2}z^2 = f(x + iy) = z_1(x, y) + iz_2(x, y)$$

which is the analytic continuation of the $\mathbb{R} \to \mathbb{R}$ parabolic function $f(x) = \frac{1}{2}x^2$.

Dimensions of a Paraboloidal Dish

The dimensions of a symmetrical paraboloidal dish are related by the equation: $4FD = R^2$, where F is the focal length, D is the depth of the dish (measured along the axis of symmetry from the vertex to the plane of the rim), and R is the radius of the rim. Of course, they must all be in the same units. If two of these three quantities are known, this equation can be used to calculate the third.

A more complex calculation is needed to find the diameter of the dish measured along its surface. This is sometimes called the "linear diameter", and equals the diameter of a flat, circular sheet of material, usually metal, which is the right size to be cut and bent to make the dish. Two intermediate results are useful in the calculation: $P = 2F$ (or the equivalent: $P = \frac{R^2}{2D}$) and $Q = \sqrt{P^2 + R^2}$, where $F, D,$ and are defined as above. The diameter of the dish, measured along the surface, is then given by: $\frac{RQ}{P} + P\ln\left(\frac{R+Q}{P}\right)$, where $\ln(x)$ means the natural logarithm of , i.e. its logarithm to base "e".

The volume of the dish, the amount of liquid it could hold if the rim were horizontal and the vertex at the bottom (e.g. the capacity of a paraboloidal wok), is given by $\frac{1}{2}\pi R^2 D$, where the symbols are defined as above. This can be compared with the formulae for the volumes of a cylinder $(\pi R^2 D)$, a hemisphere $(\frac{2}{3}\pi R^2 D$, where $D = R)$, and a cone $(\frac{1}{3}\pi R^2 D)$. Of course, πR^2 is the aperture area of the dish, the area enclosed by the rim, which is proportional to the amount of sunlight a reflector dish can intercept.

Applications

Paraboloidal mirrors are frequently used to bring parallel light to a point focus, e.g. in astronomical telescopes, or to collimate light that has originated from a source at the focus into a parallel beam, e.g. in a searchlight. The top surface of a fluid in an open-topped rotating container will form a paraboloid. This property can be used to make a liquid mirror telescope with a rotating pool of a reflective liquid, such as mercury, for the primary mirror. The same technique is used to make solid paraboloids, in rotating furnaces.

The widely-sold fried snack food Pringles potato crisps resemble a truncated hyperbolic paraboloid. According to Pringles marketing, the shape allows the snack to be securely stacked in a canister to prevent breakage during packaging and transport.

Vector Graphics

Vector graphics is the use of geometrical primitives such as points, lines, curves, and shapes or polygon(s), which are all based on mathematical expressions, to represent images in computer graphics. "Vector", in this context, implies more than a straight line.

Vector graphics are based on vectors (also called paths, or strokes) which lead through locations called control points. Each of these points has a definite position on the x and y axes of the work plan. Each point, as well, is a variety of database, including the location of the point in the work space and the direction of the vector (which is what defines the direction of the track). Each track can be assigned a colour, a shape, a thickness and also a fill. This does not affect the size of the files in a substantial way because all information resides in the structure; it describes how to draw the vector.

Overview

Computer displays are made up from grids of small rectangular cells called pixels; the term comes from "picture elements". The picture is built up from these cells. The smaller and closer the cells are together, the better the quality (resolution) of the image, but a larger file is required to store the increased pixel count. However, modern storage devices and working memory have gigabyte, even terabyte capacities, so there is less need for particularly compact forms of data.

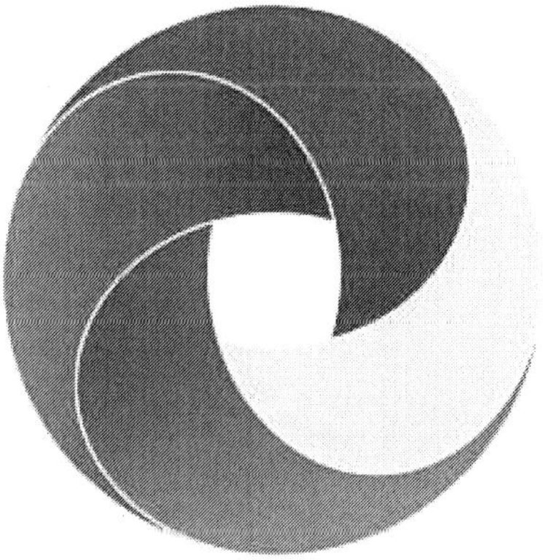

Figure: *A vector-based image of round four-colour swirl.*

Modern displays and printers are raster devices; vector formats have to be converted to raster format (bitmaps – pixel arrays) before they can be rendered (displayed or printed). The size of the bitmap/raster-format file generated by the conversion will depend on the resolution required, but the size of the vector file generating the bitmap/raster file will always remain the same. Thus, it is easy to convert from a vector file to a range of bitmap/raster file formats but it is much more difficult to go in the opposite direction, especially if subsequent editing of the vector picture is required. It might be an advantage to save an image created from a vector source file as a bitmap/raster format, because different systems have different (and incompatible) vector formats, and some might not support vector graphics at all. However, once a file is converted from the vector format, it is likely to be bigger, and it loses the advantage of scalability without loss of resolution. It will also no longer be possible to edit individual parts of the image as discrete objects. The file size of a vector graphic image depends on the number of graphic elements it contains; it is a list of descriptions.

In computer typography, modern outline fonts describe printable characters (glyphs) by cubic or quadratic mathematical curves with control points. Nevertheless, bitmap fonts are still in use. Converting outlines requires filling them in; converting to bitmaps is not trivial, because bitmaps often don't have sufficient resolution to avoid "stairstepping" ("aliasing"), especially with smaller visible character sizes. Processing outline character data in sophisticated fashion to create satisfactory bitmaps for rendering is called "hinting". Although the term implies suggestion, the process is deterministic, and done by executable code, essentially a special-purpose computer language. While automatic hinting is possible, results can be inferior to that done by experts.

Vector formats are not always appropriate in graphics work. For example, devices such as cameras and scanners produce essentially continuous-tone raster graphics that are impractical to convert into vectors, and so for this type of work, an image editor will operate on the pixels rather than on drawing objects defined by mathematical expressions. Comprehensive graphics tools will combine images from vector and raster sources, and may provide editing tools for both, since some parts of an image could come from a camera source, and others could have been drawn using vector tools.

Standards

The World Wide Web Consortium (W3C) standard for vector graphics is Scalable Vector Graphics (SVG). The standard is complex and has been relatively slow to be established at least in part owing to commercial interests. Many web browsers now have some support for rendering SVG data but full implementations of the standard are still comparatively rare.

In recent years, SVG has become a significant format that is completely independent of the resolution of the rendering device, typically a printer or display monitor. SVG files are essentially printable text that describes both straight and curved paths, as well as other attributes. Wikipedia prefers SVG for images such as simple maps, line illustrations, coats of arms, and flags, which generally are not like photographs or other continuous-tone images. Rendering SVG requires conversion to raster format at a resolution appropriate for the current task. SVG is also a format for animated graphics.

There is also a version of SVG for mobile phones. In particular, the specific format for mobile phones is called SVGT (SVG Tiny version). These images can count links and also exploit anti-aliasing. They can also be displayed as wallpaper.

Figure: *An original reference photograph before vectorization*

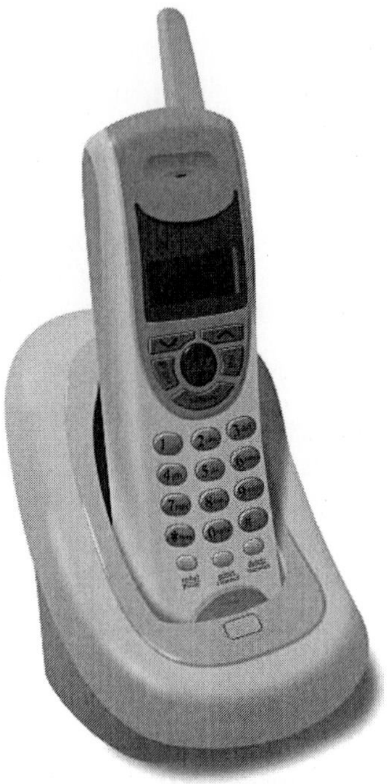

Figure: *Detail can be added or removed from vector art. Vector illustrations can have their own colour, allowing artists to achieve desired results.*

Applications

The earliest $2D$ computer graphics were all vector graphics. One of the first uses of vector graphic displays was the US SAGE air defence system. Vector graphics systems were only retired from U.S. en route air traffic control in 1999, and are likely still in use in military and specialised systems. Vector graphics were also used on the TX-2 at the MIT Lincoln Laboratory by computer graphics pioneer Ivan Sutherland to run his programme Sketchpad in 1963.

Subsequent vector graphics systems, most of which iterated through dynamically modifiable stored lists of drawing instructions, include the IBM 2250, Imlac PDS-1, and DEC GT40. There was a home gaming system that used vector graphics called Vectrex as well as various arcade games like Asteroids and Space Wars. Storage scope displays, such as the

Tektronix 4014, could display vector images but not modify them without first erasing the display. Modern vector graphics displays can sometimes be found at laser light shows, where two fast-moving X-Y mirrors position the beam to rapidly draw shapes and text as straight and curved strokes on a screen.

Vector graphics can be created in form using a pen plotter, a special type of printer that uses a series of ballpoint and/or felt-tip pens on a servo-driven mount that moves horizontally across the paper, with the plotter moving the paper back and forth through its paper path for vertical movement. Although a typical plot might easily require a few thousand paper motions, back and forth, the paper doesn't slip. In a tiny roll-fed plotter made by Alps in Japan, teeth on thin sprockets indented the paper near its edges on the first pass, and maintained registration on subsequent passes.

Some Hewlett-Packard pen plotters had two-axis pen carriers and stationary paper (plot size was limited). However, the moving-paper H-P plotters had grit wheels (akin to machine-shop grinding wheels) which, on the first pass, indented the paper surface, and collectively maintained registration.

Present-day vector graphic files such as engineering drawings are typically printed as bitmaps, after vector-to-raster conversion.

The term "vector graphics" is mainly used today in the context of two-dimensional computer graphics. It is one of several modes an artist can use to create an image on a raster display. Other modes include text, multimedia, and 3*D* rendering. Virtually all modern 3*D* rendering is done using extensions of 2*D* vector graphics techniques. Plotters used in technical drawing still draw vectors directly to paper.

Information

For example, consider a circle of radius r. The main pieces of information a programme needs in order to draw this circle are

1. an indication that what is to be drawn is a circle
2. the radius r
3. the location of the centre point of the circle

4. stroke line style and colour (possibly transparent)
5. fill style and colour (possibly transparent)

Advantages to this style of drawing over raster graphics:

- This minimal amount of information translates to a much smaller file size compared to large raster images (the size of representation does not depend on the dimensions of the object), though a vector graphic with a small file size is often said to lack detail compared with a real world photo.
- Correspondingly, one can infinitely zoom in on e.g. a circle arc, and it remains smooth. On the other hand, a polygon representing a curve will reveal being not really curved.
- On zooming in, lines and curves need not get wider proportionally. Often the width is either not increased or less than proportional. On the other hand, irregular curves represented by simple geometric shapes may be made proportionally wider when zooming in, to keep them looking smooth and not like these geometric shapes.
- The parameters of objects are stored and can be later modified. This means that moving, scaling, rotating, filling etc. doesn't degrade the quality of a drawing. Moreover, it is usual to specify the dimensions in device-independent units, which results in the best possible rasterization on raster devices.
- From a 3-D perspective, rendering shadows is also much more realistic with vector graphics, as shadows can be abstracted into the rays of light from which they are formed. This allows for photo realistic images and renderings.

Typical Primitive Objects

Any particular vector file format supports only some kinds of primitive objects. Nearly all vector file formats support simple and fast-rendering primitive objects:

- Lines, polylines and polygons
- Bézier curves and bezigons
- Circles and ellipses

Most vector file formats support

- Text (in computer font formats such as TrueType where each letter is created from Bézier curves) or quadratics.
- colour gradient
- Often, a bitmap image is considered as a primitive object. From the conceptual view, it behaves as a rectangle.

A few vector file formats support more complex objects as primitives:

- Many computer-aided design applications support splines and other curves, such as:
  * Catmull–Rom splines
  * NURBS
- iterated function systems
- superellipses and superellipsoids
- metaballs
- etc.

If an image stored in one vector file format is converted to another file format that supports all the primitive objects used in that particular image, then the conversion can be lossless.

Vector Operations

Vector graphics editors typically allow rotation, movement (without rotation), mirroring, stretching, skewing, affine transformations, changing of z-order (loosely, what's in front of what) and combination of primitives into more complex objects. More sophisticated transformations include set operations on closed shapes (union, difference, intersection, etc.).

Vector graphics are ideal for simple or composite drawings that need to be device-independent, or do not need to achieve photo-realism. For example, the PostScript and PDF page description languages use a vector graphics model.

Printing

Vector art is ideal for printing. Since the art is made from a series of mathematical curves, it will print very crisply even when resized. For instance, one can print a vector logo on a small sheet of copy paper, and then enlarge the same vector logo to billboard size and keep the same crisp quality. A low-resolution raster graphic would blur or pixelate excessively if it were enlarged from business card size to billboard size. (The precise resolution of a raster graphic necessary for high-quality results depends on the viewing distance; e.g., a billboard may still appear to be of high quality even at low resolution if the viewing distance is large enough.)

If we regard typographic characters as images, then the same considerations that we have made for graphics apply even to composition of written text for printing (typesetting). Older character sets were stored as bitmaps.

Therefore, to achieve maximum print quality they had to be used at a given resolution only; these font formats are said to be non-scalable. High quality typography is nowadays based on character drawings (fonts) which are typically stored as vector graphics, and as such are scalable to any size. Examples of these vector formats for characters are Postscript fonts and TrueType fonts.

Vector Illustration

Vector illustration is a popular technique of many digital illustrators worldwide. Some of the greatest internationally acclaimed artists in the field are Catalina Estrada, Petra Stefankova, Nathan Jurevicius, J. Otto Seibold, Matthew Inman, Leo Blanchette and others.

3D Modelling

In $3D$ computer graphics, vectorized surface representations are most common (bitmaps can be used for special purposes such as surface texturing, height-field data and bump mapping). At the low-end, simple meshes of polygons are used to represent geometric detail in applications where interactive frame rates or simplicity are important. At the high-end, where one is willing to trade-off higher rendering times for increased image quality and precision, smooth surface representations such as Bézier patches, NURBS or Subdivision surfaces are used. One can, however, achieve a smooth surface rendering from a polygonal mesh through the use of shading algorithms such as Phong and Gouraud.

Column Vector

In linear algebra, a column vector or column matrix is an m × 1 matrix, i.e. a matrix consisting of a single column of m elements.

$$\mathrm{x} = \begin{bmatrix} x_1 \\ x_2 \\ \vdots \\ x_m \end{bmatrix}$$

The transpose of a column vector is a row vector and vice versa:

$$\begin{bmatrix} x_1 \\ x_2 \\ \vdots \\ x_m \end{bmatrix}^{\mathrm{T}} = \begin{bmatrix} x_1 \; x_2 \; \dots \; x_m \end{bmatrix}$$

The set of all column vectors with a given number of elements forms a vector space which is the dual space to the set of all row vectors with that number of elements.

Notation

To simplify writing column vectors in-line with other text, sometimes they are written as row vectors with the transpose operation applied to them.

$$\mathrm{x} = \begin{bmatrix} x_1 \; x_2 \; \dots \; x_m \end{bmatrix}^{\mathrm{T}}$$

or

$$\mathrm{x} = \begin{bmatrix} x_1, x_2, \dots, x_m \end{bmatrix}^{\mathrm{T}}$$

For further simplification, some authors also use the convention of writing both column vectors and row vectors as rows, but separating row vector elements with commas and column vector elements with semicolons.

	Row vector	*Column vector*
Standard matrix notation	$\mathrm{x} = \begin{bmatrix} x_1, x_2, \dots, x_m \end{bmatrix}^{\mathrm{T}}$	$\begin{bmatrix} x_1 \\ x_2 \\ \vdots \\ x_m \end{bmatrix}$ or $\begin{bmatrix} x_1 \; x_2 \; \dots \; x_m \end{bmatrix}^{\mathrm{T}}$
Alternative notation 1	$\begin{bmatrix} x_1, x_2, \dots, x_m \end{bmatrix}$	$\begin{bmatrix} x_1, x_2, \dots, x_m \end{bmatrix}^{\mathrm{T}}$
Alternative notation 2	$\begin{bmatrix} x_1, x_2, \dots, x_m \end{bmatrix}$	$\begin{bmatrix} x_1 ; x_2 ; \dots ; x_m \end{bmatrix}$

Operations

- Matrix multiplication involves the action of multiplying each row vector of one matrix by each column vector of another matrix.
- The dot product of two vectors a and b is equivalent to multiplying the row vector representation of a by the column vector representation of b:

$$\mathrm{a}\cdot\mathrm{b} = \mathrm{a}^{\mathrm{T}}\mathrm{b} = \begin{bmatrix} a_1 & a_2 & a_3 \end{bmatrix} \begin{bmatrix} b_1 \\ b_2 \\ b_3 \end{bmatrix}.$$

Coordinate Vector

In linear algebra, the coordinate representation or coordinate vector of a vector is the unique tuple of numbers that describes the vector in terms of a particular ordered basis. Coordinates are always specified relative to an ordered basis. Bases and their associated coordinate representations let one realize vector spaces and linear transformations concretely as column vectors, row vectors, and matrices, hence are useful in calculations. The idea of a coordinate vector can also be used for infinite dimensional vector spaces, as addressed below.

Definition

Let V be a vector space of dimension n over a field F and let

$$B = \{b_1, b_2, \ldots, b_n\}$$

be an ordered basis for V. Then for every $v \in V$ there is a unique linear combination of the basis vectors that equals v:

$$v = \alpha_1 b_1 + \alpha_2 b_2 + \cdots + \alpha_n b_n.$$

The coordinate vector of v relative to B is the sequence of coordinates

$$[v]_B = (\alpha_1, \alpha_2, \cdots, \alpha_n).$$

This is also called the representation of v with respect of B, or the B representation of v. The α-s are called the coordinates of v. The order of the basis becomes important here, since it determines the order in which the coefficients are listed in the coordinate vector.

Coordinate vectors of finite dimensional vector spaces can be represented as columns or row vectors. In the above notation, one can write

$$[v]_B = \begin{bmatrix} \alpha_1 \\ \vdots \\ \alpha_n \end{bmatrix}$$

or

$$[v]_B = [\alpha_1, \ldots, \alpha_n].$$

The Standard Representation

We can mechanize the above transformation by defining a function ϕ_B, called the standard representation of V with respect to B, that takes every vector to its coordinate representation: $\phi_B(v) = [v]_B$.

Then ϕ_B is a linear transformation from V to F^n. In fact, it is an isomorphism, and its inverse $\phi_B^{-1} : F^n \to V$ is simply

$$\phi_B^{-1}(\alpha_1, \ldots, \alpha_n) = \alpha_1 b_1 + \cdots + \alpha_n b_n.$$

Alternatively, we could have defined ϕ_B^{-1} to be the above function from the beginning, realized that ϕ_B^{-1} is an isomorphism, and defined ϕ_B to be its inverse.

Example 1

Let $P4$ be the space of all the algebraic polynomials in degree less than 4 (i.e. the highest exponent of x can be 3). This space is linear and spanned by the following polynomials:

$$B_P = \{1, x, x^2, x^3\}$$

matching

$$1 := \begin{bmatrix} 1 \\ 0 \\ 0 \\ 0 \end{bmatrix} \;;\; x := \begin{bmatrix} 0 \\ 1 \\ 0 \\ 0 \end{bmatrix} \;;\; x^2 := \begin{bmatrix} 0 \\ 0 \\ 1 \\ 0 \end{bmatrix} \;;\; x^3 := \begin{bmatrix} 0 \\ 0 \\ 0 \\ 1 \end{bmatrix}$$

then the corresponding coordinate vector to the polynomial

$$p(x) = a_0 + a_1 x + a_2 x^2 + a_3 x^3 \text{ is } \begin{bmatrix} a_0 \\ a_1 \\ a_2 \\ a_3 \end{bmatrix}.$$

According to that representation, the differentiation operator d/dx which we shall mark D will be represented by the following matrix:

$$Dp(x) = P'(x) \quad ; \quad [D] = \begin{bmatrix} 0 & 1 & 0 & 0 \\ 0 & 0 & 2 & 0 \\ 0 & 0 & 0 & 3 \\ 0 & 0 & 0 & 0 \end{bmatrix}$$

Using that method it is easy to explore the properties of the operator: such as invertibility, hermitian or anti-hermitian or none, spectrum and eigenvalues and more.

Example 2

The Pauli matrices which represent the spin operator when transforming the spin eigenstates into vector coordinates.

Basis Transformation Matrix

Let B and C be two different bases of a vector space V, and let us mark with $[M]_C^B$ the matrix which has columns consisting of the C representation of basis vectors b_1, b_2, ..., b_n:

$$[M]_C^B = [\ [b_1]_C \ \cdots \ [b_n]_C\ ]$$

This matrix is referred to as the basis transformation matrix from B to C, and can be used for transforming any vector v from a B representation to a C representation, according to the following theorem:

$$[v]_C = [M]_C^B [v]_B.$$

If E is the standard basis, the transformation from B to E can be represented with the following simplified notation:

$$v = [M]^B [v]_B.$$

where

$$v = [v]_E, \text{ and } [M]^B = [M]_E^B.$$

Corollary

The matrix M is an invertible matrix and M^{-1} is the basis transformation matrix from C to B. In other words,

$$[M]_C^B [M]_B^C = [M]_C^C = \text{Id}$$

$$[M]_B^C [M]_C^B = [M]_B^B = \text{Id}$$

Remarks

1. The basis transformation matrix can be regarded as an automorphism over V.
2. In order to easily remember the theorem $[v]_C = [M]_C^B [v]_B$, notice that M 's superscript and v 's subscript indices are "cancelling" each other and M 's subscript becomes v 's new subscript. This "cancelling" of indices is not a real cancelling but rather a convenient and intuitively appealing, although mathematically incorrect, manipulation of symbols, permitted by an appropriately chosen notation.

Infinite Dimensional Vector Spaces

Suppose V is an infinite dimensional vector space over a field F. If the dimension is κ, then there is some basis of κ elements for V. After an order is chosen, the basis can be considered an ordered basis.

The elements of V are finite linear combinations of elements in the basis, which give rise to unique coordinate representations exactly as described before.

The only change is that the indexing set for the coordinates is not finite. Since a given vector v is a finite linear combination of basis elements, the only nonzero entries of the coordinate vector for v will be the nonzero coefficients of the linear combination representing v. Thus the coordinate vector for v is zero except in finitely many entries.

The linear transformations between (possibly) infinite dimensional vector spaces can be modelled, analogously to the finite dimensional case, with infinite matrices. The special case of the transformations from V into V is described in the full linear ring article.

Dimension (Vector Space)

In mathematics, the dimension of a vector space V is the cardinality (i.e. the number of vectors) of a basis of V.

For every vector space there exists a basis (if one assumes the axiom of choice), and all bases of a vector space have equal cardinality; as a result the dimension of a vector space is uniquely defined. We say V is finite-dimensional if the dimension of V is finite.

The dimension of the vector space V over the field F can be written as $\dim_F(V)$ or as $[V : F]$, read "dimension of V over F". When F can be inferred from context, often just $\dim(V)$ is written.

Examples

The vector space R^3 has

$$\left\{ \begin{pmatrix} 1 \\ 0 \\ 0 \end{pmatrix}, \begin{pmatrix} 0 \\ 1 \\ 0 \end{pmatrix}, \begin{pmatrix} 0 \\ 0 \\ 1 \end{pmatrix} \right\}$$

as a basis, and therefore we have $\dim_R(R^3) = 3$. More generally, $\dim_R(R^n) = n$, and even more generally, $\dim_F(F^n) = n$ for any field F.

The complex numbers C are both a real and complex vector space; we have $\dim_R(C) = 2$ and $\dim_C(C) = 1$. So the dimension depends on the base field.

The only vector space with dimension 0 is $\{0\}$, the vector space consisting only of its zero element.

Facts

If W is a linear subspace of V, then $\dim(W) \leq \dim(V)$.

To show that two finite-dimensional vector spaces are equal, one often uses the following criterion: if V is a finite-dimensional vector space and W is a linear subspace of V with $\dim(W) = \dim(V)$, then $W = V$.

R^n has the standard basis $\{e_1, ..., e_n\}$, where e_i is the i-th column of the corresponding identity matrix. Therefore R^n has dimension n. Any two vector spaces over F having the same dimension are isomorphic. Any bijective map between their bases can be uniquely extended to a bijective linear map between the vector spaces. If *B* is some set, a vector space with dimension $|B|$ over F can be constructed as follows: take the set $F^{(B)}$ of all functions $f: B \to F$ such that $f(b) = 0$ for all but finitely many *b* in *B*. These functions can be added and multiplied with elements of F, and we obtain the desired F-vector space.

An important result about dimensions is given by the rank–nullity theorem for linear maps. If F/K is a field extension, then *F* is in particular a vector space over K. Furthermore, every *F*-vector space *V* is also a *K*-vector space. The dimensions are related by the formula

$$\dim_K(V) = \dim_K(F) \dim_F(V).$$

In particular, every complex vector space of dimension n is a real vector space of dimension 2n.

Some simple formulae relate the dimension of a vector space with the cardinality of the base field and the cardinality of the space itself. If *V* is a vector space over a field *F* then, denoting the dimension of *V* by dim*V*, we have:

If dim *V* is finite, then $|V| = |F|^{\dim V}$.

If dim *V* is infinite, then $|V| = \max(|F|, \dim V)$.

Generalizations

One can see a vector space as a particular case of a matroid, and in the latter there is a well-defined notion of dimension. The length of a module and the rank of an abelian group both have several properties similar to the dimension of vector spaces.

The Krull dimension of a commutative ring, named after Wolfgang Krull (1899–1971), is defined to be the maximal number of strict inclusions in an increasing chain of prime ideals in the ring.

Trace

The dimension of a vector space may alternatively be characterized as the trace of the identity operator. For instance,

$\text{tr id}_{\mathrm{R}^2} = \text{tr}\begin{pmatrix} 1 & 0 \\ 0 & 1 \end{pmatrix} = 1 + 1 = 2$. This begs the definition of trace, but allows useful generalizations.

Firstly, it allows one to define a notion of dimension when one has a trace but no natural sense of basis. For example, one may have an algebra A with maps $\eta : K \to A$ (the inclusion of scalars, called the unit) and a map $\epsilon : A \to K$ (corresponding to trace, called the counit). The composition $\epsilon^\circ \eta : K \to K$ is a scalar (being a linear operator on a 1-dimensional space) corresponds to "trace of identity", and gives a notion of dimension for an abstract algebra. In practice, in bialgebras one requires that this map be the identity, which can be obtained by normalizing the counit by dividing by dimension ($\epsilon := \frac{1}{n}\text{tr}$), so in these cases the normalizing constant corresponds to dimension.

Alternatively, one may be able to take the trace of operators on an infinite-dimensional space; in this case a (finite) trace is defined, even though no (finite) dimension exists, and gives a notion of "dimension of the operator". These fall under the rubric of "trace class operators" on a Hilbert space, or more generally nuclear operators on a Banach space.

A subtler generalization is to consider the trace of a family of operators as a kind of "twisted" dimension. This occurs significantly in representation theory, where the character of a representation is the trace of the representation, hence a scalar-valued function on a group $\chi : G \to K$, whose value on the identity $1 \in G$ is the dimension of the representation, as a representation sends the identity in the group to the identity matrix: $\chi(1_G) = \text{tr } I_V = \dim V$. One can view the other values $\chi(g)$ of the character as "twisted" dimensions, and find analogs or generalizations of statements about dimensions to statements about characters or representations. A sophisticated example of this occurs in the theory of monstrous moonshine: the j-invariant is the graded dimension of an infinite-dimensional graded representation of the Monster group, and replacing the dimension with the character gives the McKay–Thompson series for each element of the Monster group.

Direction Vector

In mathematics, a direction vector that describes a line segment D is any vector

$$\overrightarrow{AB}$$

where $A$and $B$are two distinct points on the line D. If v is a direction vector for D, so is kv for any nonzero scalar k; and these are in fact all of the direction vectors for the line D. Under some definitions, the direction vector is required to be a unit vector, in which case each line has exactly two direction vectors, which are negatives of each other (equal in magnitude, opposite in direction).

Direction Vector for a Line in R^2

Any line in two-dimensional Euclidean space can be described as the set of solutions to an equation of the form

$$ax + by + c = 0$$

where a, b, c are real numbers. Then one direction vector of (D)is $(-b,a)$. Any multiple of $(-b,a)$ is also a direction vector.

For example, suppose the equation of a line is $3x + 2y + 15 = 0$. Then $(-2,3)$, $(-4,6)$, and $(2,-3)$ are all direction vectors for this line.

Parametric Equation for a Line

In Euclidean space (any number of dimensions), given a point a and a nonzero vector v, a line is defined parametrically by (a+tv), where the parameter t varies between $-\infty$ and $+\infty$. This line has v as a direction vector.

Direction Cosine

In analytic geometry, the direction cosines (or directional cosines) of a vector are the cosines of the angles between the vector and the three coordinate axes. Or equivalently it is the component contributions of the basis to the unit vector.

Mathematical Formula

If v is a vector

$$\mathrm{v} = v_1\hat{\mathrm{x}} + v_2\hat{\mathrm{y}} + v_3\hat{\mathrm{z}}$$

where $\hat{\mathrm{x}},\hat{\mathrm{y}},\hat{\mathrm{z}}$ are basis, then the direction cosines are

$$\alpha = \cos a = \frac{\mathbf{v}\cdot\hat{\mathbf{x}}}{\|\mathbf{v}\|} = \frac{v_1}{\sqrt{v_1^2 + v_2^2 + v_3^2}},$$

$$\beta = \cos b = \frac{\mathbf{v}\cdot\hat{\mathbf{y}}}{\|\mathbf{v}\|} = \frac{v_2}{\sqrt{v_1^2 + v_2^2 + v_3^2}},$$

$$\gamma = \cos c = \frac{\mathbf{v}\cdot\hat{\mathbf{z}}}{\|\mathbf{v}\|} = \frac{v_3}{\sqrt{v_1^2 + v_2^2 + v_3^2}}.$$

Note that

$$\cos^2 a + \cos^2 b + \cos^2 c = 1$$

and

(α,β,γ) are the Cartesian coordinates of the unit vector $\hat{\mathrm{v}}$.

Here, α, β and γ are the Direction cosines and a, b and c are the Direction angles of the vector v.

The Direction angles a, b and c are acute, i.e., $0 \le a \le \pi$, $0 \le b \le \pi$ and $0 \le c \le \pi$ and they denote the angles formed between v and the unit coordinate vectors, $\hat{\imath}$, $\hat{\jmath}$ and $\hat{k}$.

General Meaning

More generally, direction cosine refers to the cosine of the angle between any two vectors. They are useful for forming direction cosine matrices that express one set of orthonormal basis vectors in terms of another set, or for expressing a known vector in a different basis.

Chapter 2

Eccentricity Vector

In celestial mechanics, the eccentricity vector of a Kepler orbit is the vector that points towards the periapsis and has a magnitude equal to the orbit's scalar eccentricity.

The magnitude is unitless. For Kepler orbits the eccentricity vector is a constant of motion.

Its main use is in the analysis of almost circular orbits, as perturbing (non-Keplerian) forces on an actual orbit will cause the osculating eccentricity vector to change continuously.

For the eccentricity and argument of periapsis parameters, eccentricity zero (circular orbit) corresponds to a singularity.

Calculation

The eccentricity vector e can be calculated

$$e = \frac{v \times h}{\mu} - \frac{r}{|r|}$$

where:

- vis velocity vector
- his specific angular momentum vector (equal to $r \times v$)
- ris position vector
- μ is standard gravitational parameter

or equivalently:

$$e = \frac{|v|^2 r}{\mu} - \frac{(r \cdot v)v}{\mu} - \frac{r}{|r|}$$

Derivation

The velocity vector $v$can be derived from position vector r by differentiation with respect to time:

$$\mathrm{v} = \frac{dr}{dt}$$

An object's state vector can be used to compute its classical or Keplerian orbital elements and vice versa. Each representation has its advantages. The elements are more descriptive of the size, shape and orientation of an orbit, and may be used to quickly and easily estimate the object's state at any arbitrary time provided its motion is accurately modelled by the two-body problem with only small perturbations.

On the other hand, the state vector is more directly useful in a numerical integration that accounts for significant, arbitrary, time-varying forces such as drag, thrust and gravitational perturbations from third bodies as well as the gravity of the primary body.

The state vectors (r and v) can be easily used to compute the angular momentum vector as $\mathrm{h} = \mathrm{r} \times \mathrm{v}$.

Because even satellites in low earth orbit experience significant perturbations (primarily from the non-spherical shape of the earth), the Keplerian elements computed from the state vector at any moment are only valid at that time. Such element sets are known as osculating elements because they coincide with the actual orbit only at that moment.

Probability Vector

In mathematics and statistics, a probability vector or stochastic vector is a vector with non-negative entries that add up to one.

The positions (indices) of a probability vector represent the possible outcomes of a discrete random variable, and the vector gives us the probability mass function of that random variable, which is the standard way of characterizing a discrete probability distribution.

Here are some examples of probability vectors. The vectors can be either columns or rows.

$$x_0 = \begin{bmatrix} 0.5 \\ 0.25 \\ 0.25 \end{bmatrix}, x_1 = \begin{bmatrix} 0 \\ 1 \\ 0 \end{bmatrix}, x_2 = \begin{bmatrix} 0.65 & 0.35 \end{bmatrix}, x_3 = \begin{bmatrix} 0.3 & 0.5 & 0.07 & 0.1 & 0.03 \end{bmatrix}.$$

Writing out the vector components of a vector $p$as

$$p = \begin{bmatrix} p_1 \\ p_2 \\ \vdots \\ p_n \end{bmatrix} \quad \text{or} \quad p = \begin{bmatrix} p_1 & p_2 & \cdots & p_n \end{bmatrix}$$

the vector components must sum to one:

$$\sum_{i=1}^{n} p_i = 1$$

One also has the requirement that each individual component must have a probability between zero and one:

$$0 \le p_i \le 1$$

for all i. These two requirements show that stochastic vectors have a geometric interpretation: A stochastic vector is a point on the "far face" of a standard orthogonal simplex. That is, a stochastic vector uniquely identifies a point on the face opposite of the orthogonal corner of the standard simplex.

Pseudovector

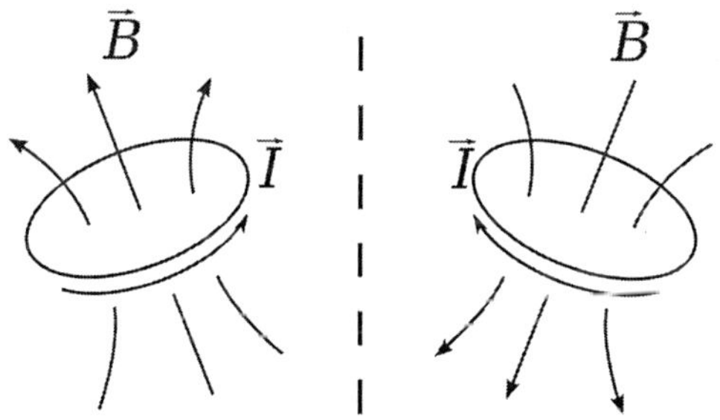

Figure: *A loop of wire (black), carrying a current, creates a magnetic field (blue). If the position and current of the wire are reflected across the dotted line, the magnetic field it generates would not be reflected: Instead, it would be reflected and reversed. The position of the wire and its current are vectors, but the magnetic field is a pseudovector.*

In physics and mathematics, a pseudovector (or axial vector) is a quantity that transforms like a vector under a proper rotation, but gains an additional sign flip under an improper rotation such as a reflection. Geometrically it is the opposite, of equal magnitude but in the opposite direction, of its mirror image. This is as opposed to a true or polar vector (more formally, a contravariant vector), which on reflection matches its mirror image.

In three dimensions the pseudovector p is associated with the cross product of two polar vectors a and b:

$$p = a \times b.$$

The vector p calculated this way is a pseudovector. One example is the normal to an oriented plane. An oriented plane can be defined by two non-parallel vectors, a and b, which can be said to span the plane. The vector a × b is a normal to the plane (there are two normals, one on each side — the right-hand rule will determine which), and is a pseudovector. This has consequences in computer graphics where it has to be considered when transforming surface normals.

A number of quantities in physics behave as pseudovectors rather than polar vectors, including magnetic field and angular velocity. In mathematics pseudovectors are equivalent to three dimensional bivectors, from which the transformation rules of pseudovectors can be derived. More generally in n-dimensional geometric algebra pseudovectors are the elements of the algebra with dimension $n - 1$, written $\wedge^{n-1}R^n$. The label 'pseudo' can be further generalized to pseudoscalars and pseudotensors, both of which gain an extra sign flip under improper rotations compared to a true scalar or tensor.

Physical Examples

Physical examples of pseudovectors include the magnetic field, torque, vorticity, and the angular momentum.

Often, the distinction between vectors and pseudovectors is overlooked, but it becomes important in understanding and exploiting the effect of symmetry on the solution to physical systems. For example, consider the case of an electrical current loop in the $z = 0$ plane, which has a magnetic field at $z = 0$ that is oriented in the z direction. This system is symmetric (invariant) under mirror reflections through the plane (an improper rotation), so the magnetic field should be unchanged by the reflection. But reflecting the magnetic field through that plane naively appears to change its sign if it is viewed as a vector field—this contradiction is resolved by realizing that the mirror reflection of the field induces an extra sign flip because of its pseudovector nature, so the mirror flip in the end leaves the magnetic field unchanged as expected.

As another example, consider the pseudovector angular momentum $L = r \times p$. Driving in a car, and looking forward, each of the wheels has an angular momentum vector pointing to the left. If the world is reflected in a mirror which switches the left and right side of the car, the "reflection" of this angular momentum "vector" (viewed as an ordinary vector) points to the right, but the actual angular momentum vector of the wheel still points to the left, corresponding to the extra minus sign in the reflection of a pseudovector. This reflects the fact that the wheels are still turning forward. In comparison, the behaviour of a regular vector, such as the position of the car, is quite different.

To the extent that physical laws would be the same if the universe were reflected in a mirror (equivalently, invariant under parity), the sum of a vector and a pseudovector is not meaningful. However, the weak force, which governs beta decay, does depend on the chirality of the universe, and in this case pseudovectors and vectors are added.

Details

The definition of a "vector" in physics (including both polar vectors and pseudovectors) is more specific than the mathematical definition of "vector" (namely, any element of an abstract vector space). Under the physics definition, a "vector" is required to have components that "transform" in a certain way under a proper rotation: In particular, if everything in the universe were rotated, the vector would rotate in exactly the same way. (The coordinate system is fixed in this discussion; in other words this is the perspective of active transformations.) Mathematically, if everything in the universe undergoes a rotation described by a rotation matrix R, so that a displacement vector x is transformed to $x' = Rx$, then any "vector" v must be similarly transformed to v' $= Rv$. This important requirement is what distinguishes a vector (which might be composed of, for example, the x, y, and z-components of velocity) from any other triplet of physical quantities (For example, the length, width, and height of a rectangular box cannot be considered the three components of a vector, since rotating the box does not appropriately transform these three components.)

(In the language of differential geometry, this requirement is equivalent to defining a vector to be a tensor of contravariant rank one.)

The discussion so far only relates to proper rotations, i.e. rotations about an axis. However, one can also consider improper rotations, i.e. a mirror-reflection possibly followed by a proper rotation. (One example of an improper rotation is inversion.) Suppose everything in the universe undergoes an improper rotation described by the rotation matrix R, so that a position vector x is transformed to $x' = Rx$. If the vector v is a polar vector, it will be transformed to v' = Rv. If it is a pseudovector, it will be transformed to $v' = -Rv$.

The transformation rules for polar vectors and pseudovectors can be compactly stated as

$$\mathrm{v}' = R\mathrm{v} \text{ (polar vector)}$$

$$\mathrm{v}' = (\det R)(R\mathrm{v}) \text{ (pseudovector)}$$

where the symbols are as described above, and the rotation matrix R can be either proper or improper. The symbol det denotes determinant; this formula works because the determinant of proper and improper rotation matrices are +1 and -1, respectively.

Behaviour Under Addition, Subtraction, Scalar Multiplication

Suppose v_1 and v_2 are known pseudovectors, and v_3 is defined to be their sum, $v_3=v_1+v_2$. If the universe is transformed by a rotation matrix R, then v_3 is transformed to

$$\mathrm{v}_3' = \mathrm{v}_1' + \mathrm{v}_2' = (\det R)(R\mathrm{v}_1) + (\det R)(R\mathrm{v}_2) = (\det R)(R(\mathrm{v}_1 + \mathrm{v}_2)) = (\det R)(R\mathrm{v}_3).$$

So v_3 is also a pseudovector. Similarly one can show that the difference between two pseudovectors is a pseudovector, that the sum or difference of two polar vectors is a polar vector, that multiplying a polar vector by any real number yields another polar vector, and that multiplying a pseudovector by any real number yields another pseudovector.

On the other hand, suppose v_1 is known to be a polar vector, v_2 is known to be a pseudovector, and v_3 is defined to be their sum, $v_3=v_1+v_2$. If the universe is transformed by a rotation matrix R, then v_3 is transformed to

$$v_3' = v_1' + v_2' = (Rv_1) + (\det R)(Rv_2) = R(v_1 + (\det R)v_2).$$

Therefore, v_3 is neither a polar vector nor a pseudovector. For an improper rotation, v_3 does not in general even keep the same magnitude:

$$|v_3| = |v_1 + v_2|, \text{but } |v_3'| = |v_1' - v_2'|.$$

If the magnitude of v_3 were to describe a measurable physical quantity, that would mean that the laws of physics would not appear the same if the universe was viewed in a mirror. In fact, this is exactly what happens in the weak interaction: Certain radioactive decays treat "left" and "right" differently, a phenomenon which can be traced to the summation of a polar vector with a pseudovector in the underlying theory.

Behaviour Under Cross Products

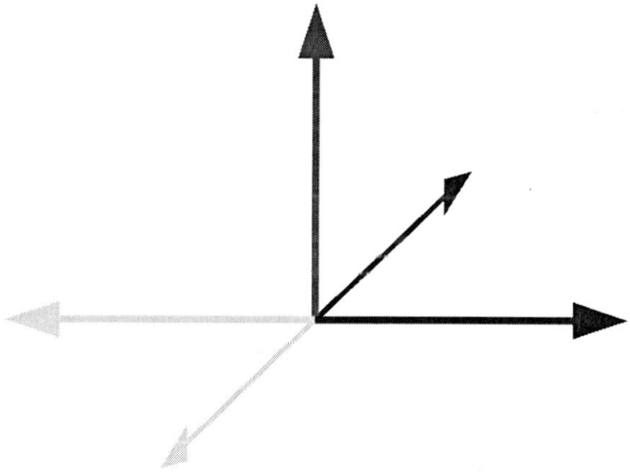

Figure: *Under inversion the two vectors change sign, but their cross product is invariant [black are the two original vectors, grey are the inverted vectors, and red is their mutual cross product].*

For a rotation matrix R, either proper or improper, the following mathematical equation is always true:

$$(Rv_1)\times(Rv_2) = (\det R)(R(v_1 \times v_2)),$$

where v_1 and v_2 are any three-dimensional vectors. (This equation can be proven either through a geometric argument or through an algebraic calculation, and is well known.)

Suppose v_1 and v_2 are known polar vectors, and v_3 is defined to be their cross product, $v_3=v_1\times v_2$. If the universe is transformed by a rotation matrix R, then v_3 is transformed to

$$v_3' = v_1' \times v_2' = (Rv_1)\times(Rv_2) = (\det R)(R(v_1 \times v_2)) = (\det R)(Rv_3).$$

So v_3 is a pseudovector. Similarly, one can show:

- polar vector × polar vector = pseudovector
- pseudovector × pseudovector = pseudovector
- polar vector × pseudovector = polar vector
- pseudovector × polar vector = polar vector

Examples

From the definition, it is clear that a displacement vector is a polar vector. The velocity vector is a displacement vector (a polar vector) divided by time (a scalar), so is also a polar vector. Likewise, the momentum vector is the velocity vector (a polar vector) times mass (a scalar), so is a polar vector. Angular momentum is the cross product of a displacement (a polar vector) and momentum (a polar vector), and is therefore a pseudovector. Continuing this way, it is straightforward to classify any vector as either a pseudovector or polar vector.

The Right-hand Rule

Above, pseudovectors have been discussed using active transformations. An alternate approach, more along the lines of passive transformations, is to keep the universe fixed, but switch "right-hand rule" with "left-hand rule" and vice-versa everywhere in physics, in particular in the definition of the cross product. Any polar vector (e.g., a translation vector) would be unchanged, but pseudovectors (e.g., the magnetic field vector at a point) would switch signs. Nevertheless, there would be no physical consequences, apart from in the parity-violating phenomena such as certain radioactive decays.

Geometric Algebra

In geometric algebra the basic elements are vectors, and these are used to build a hierarchy of elements using the definitions of products in this algebra. In particular, the algebra builds pseudovectors from vectors.

The basic multiplication in the geometric algebra is the geometric product, denoted by simply juxtaposing two vectors as in ab. This product is expressed as:

$$\mathrm{ab} = \mathrm{a} \cdot \mathrm{b} + \mathrm{a} \wedge \mathrm{b} ,$$

where the leading term is the customary vector dot product and the second term is called the wedge product. Using the postulates of the algebra, all combinations of dot and wedge products can be evaluated. A terminology to describe the various combinations is provided. For example, a multivector is a summation of k-fold wedge products of various k-values. A k-fold wedge product also is referred to as a k-blade.

In the present context the pseudovector is one of these combinations. This term is attached to a different mulitvector depending upon the dimensions of the space (that is, the number of linearly independent vectors in the space). In three dimensions, the most general 2-blade or bivector can be expressed as a single wedge product and is a pseudovector. In four dimensions, however, the pseudovectors are trivectors. In general, it is a (n - 1)-blade, where n is the dimension of the space and algebra. An n-dimensional space has n vectors and also n pseudovectors. Each pseudovector is formed from the outer (wedge) product of all but one of the n vectors. For instance, in four dimensions where the vectors are: $\{e_1, e_2, e_3, e_4\}$, the pseudovectors can be written as: $\{e_{234}, e_{134}, e_{124}, e_{123}\}$.

Transformations in Three Dimensions

The transformation properties of the pseudovector in three dimensions has been compared to that of the vector cross product by Baylis. He says: "The terms axial vector and pseudovector are often treated as synonymous, but it is quite useful to be able to distinguish a bivector from its dual." To paraphrase Baylis: Given two polar vectors (that is, true vectors) a and b in three dimensions, the cross product composed from a and b is the vector normal to their plane given by $c = a \times b$. Given a set of right-handed orthonormal basis vectors $\{ e_{\ell} \}$, the cross product is expressed in terms of its components as:

$$a \times b = (a^2b^3 - a^3b^2)e_1 + (a^3b^1 - a^1b^3)e_2 + (a^1b^2 - a^2b^1)e_3,$$

where superscripts label vector components. On the other hand, the plane of the two vectors is represented by the exterior product or wedge product, denoted by $a \wedge b$. In this context of geometric algebra, this bivector is called a pseudovector, and is the dual of the cross product. The dual of e_1 is introduced as e_{23} a+ $e_2e_3 = e_2 \wedge e_3$, and so forth. That is, the dual of e_1 is the subspace perpendicular to e_1, namely the subspace spanned by e_2 and e_3. With this understanding,

$$a \wedge b = (a^2b^3 - a^3b^2)e_{23} + (a^3b^1 - a^1b^3)e_{31} + (a^1b^2 - a^2b^1)e_{12} .$$

Comparison shows that the cross product and wedge product are related by:

$$a \wedge b = i\, a \times b ,$$

where $i = e_1 \wedge e_2 \wedge e_3$ is called the unit pseudoscalar. It has the property:

$$i^2 = -1 .$$

Using the above relations, it is seen that if the vectors a and b are inverted by changing the signs of their components while leaving the basis vectors fixed, both the pseudovector and the cross product are invariant. On the other hand, if the components are fixed and the basis vectors e_{ℓ} are inverted, then the pseudovector is invariant, but the cross product changes sign. This behaviour of cross products is consistent with their definition as vector-like elements that change sign under transformation from a right-handed to a left-handed coordinate system, unlike polar vectors.

Note on Usage

As an aside, it may be noted that not all authors in the field of geometric algebra use the term pseudovector, and some authors follow the terminology that does not distinguish between the pseudovector and the cross product. However, because the cross product does not generalize beyond three dimensions, the notion of pseudovector based upon the cross product also cannot be extended to higher dimensions. The pseudovector as the (n–1)-blade of an n-dimensional space is not so restricted. Another important note is that pseudovectors, despite their name, are "vectors" in the common mathematical sense, i.e. elements of a vector space. The idea that "a pseudovector is different from a vector" is only true with a different and more specific definition of the term "vector" as discussed above.

Laplace–Runge–Lenz Vector

In classical mechanics, the Laplace–Runge–Lenz vector (or simply the LRL vector) is a vector used chiefly to describe the shape and orientation of the orbit of one astronomical body around another, such as a planet revolving around a star. For two bodies interacting by Newtonian gravity, the LRL vector is a constant of motion, meaning that it is the same no matter where it is calculated on the orbit; equivalently, the LRL vector is said to be conserved. More generally, the LRL vector is conserved in all problems in which two bodies interact by a central force that varies as the inverse square of the distance between them; such problems are called Kepler problems.

The hydrogen atom is a Kepler problem, since it comprises two charged particles interacting by Coulomb's law of electrostatics, another inverse square central force. The LRL vector was essential in the first quantum mechanical derivation of the spectrum of the hydrogen atom, before the development of the Schrödinger equation. However, this approach is rarely used today.

In classical and quantum mechanics, conserved quantities generally correspond to a symmetry of the system. The conservation of the LRL vector corresponds to an unusual symmetry; the Kepler problem is mathematically equivalent to a particle moving freely on the surface of a four-dimensional (hyper-)sphere, so that the whole problem is symmetric under certain rotations of the four-dimensional space. This higher symmetry results from two properties of the Kepler problem: the velocity vector always moves in a perfect circle and, for a given total energy, all such velocity circles intersect each other in the same two points. The Laplace–Runge–Lenz vector is named after Pierre-Simon de Laplace, Carl Runge and Wilhelm

Lenz. It is also known as the Laplace vector, the Runge–Lenz vector and the Lenz vector. Ironically, none of those scientists discovered it. The LRL vector has been re-discovered several times and is also equivalent to the dimensionless eccentricity vector of celestial mechanics. Various generalizations of the LRL vector have been defined, which incorporate the effects of special relativity, electromagnetic fields and even different types of central forces.

Context

A single particle moving under any conservative central force has at least four constants of motion, the total energy E and the three Cartesian components of the angular momentum vector L. The particle's orbit is confined to a plane defined by the particle's initial momentum p (or, equivalently, its velocity v) and the vector r between the particle and the centre of force.

The Laplace–Runge–Lenz vector (LRL vector) A always lies in the plane of motion for any central force. However, A is constant only for an inverse-square central force. For most central forces, however, this vector A is not constant, but changes in both length and direction; if the central force is approximately an inverse-square law, the vector A is approximately constant in length, but slowly rotates its direction. A generalized conserved LRL vector $\mathcal{A}$ can be defined for all central forces, but this generalized vector is a complicated function of position, and usually not expressible in closed form.

The plane of motion is perpendicular to the angular momentum vector L, which is constant; this may be expressed mathematically by the vector dot product equation $r \cdot L = 0$; likewise, since A lies in that plane, $A \cdot L = 0$.

The LRL vector differs from other conserved quantities in the following property. Whereas for typical conserved quantities, there is a corresponding cyclic coordinate in the three-dimensional Lagrangian of the system, there does not exist such a coordinate for the LRL vector. Thus, the conservation of the LRL vector must be derived directly, e.g., by the method of Poisson brackets, as described below. Conserved quantities of this kind are called "dynamic", in contrast to the usual "geometric" conservation laws, e.g., that of the angular momentum.

History of Rediscovery

The LRL vector A is a constant of motion of the important Kepler problem, and is useful in describing astronomical orbits, such as the motion of the planets.

Nevertheless, it has never been well known among physicists, possibly because it is less intuitive than momentum and angular momentum.

Consequently, it has been rediscovered independently several times over the last three centuries.

Jakob Hermann was the first to show that A is conserved for a special case of the inverse-square central force, and worked out its connection to the eccentricity of the orbital ellipse.

Hermann's work was generalized to its modern form by Johann Bernoulli in 1710.

At the end of the century, Pierre-Simon de Laplace rediscovered the conservation of A, deriving it analytically, rather than geometrically.

In the middle of the nineteenth century, William Rowan Hamilton derived the equivalent eccentricity vector defined below, using it to show that the momentum vector p moves on a circle for motion under an inverse-square central force.

At the beginning of the twentieth century, Josiah Willard Gibbs derived the same vector by vector analysis. Gibbs' derivation was used as an example by Carle Runge in a popular German textbook on vectors, which was referenced by Wilhelm Lenz in his paper on the (old) quantum mechanical treatment of the hydrogen atom.

In 1926, the vector was used by Wolfgang Pauli to derive the spectrum of hydrogen using modern quantum mechanics, but not the Schrödinger equation; after Pauli's publication, it became known mainly as the Runge–Lenz vector.

Mathematical Definition

For a single particle acted on by an inverse-square central force described by the equation $\mathrm{F}(r) = \frac{-k}{r^2}\hat{\mathrm{r}}$, the LRL vector A is defined mathematically by the formula

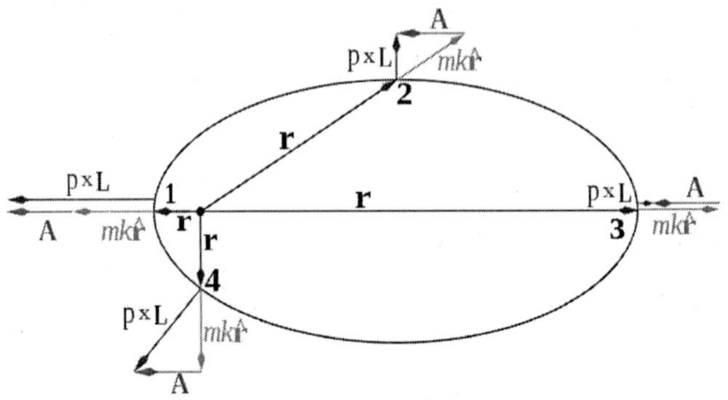

Figure: *The LRL vector A (shown in red) at four points (labelled 1, 2, 3 and 4) on the elliptical orbit of a bound point particle moving under an inverse-square central force. The centre of attraction is shown as a small black circle from which the position vectors (likewise black) emanate. The angular momentum vector L is perpendicular to the orbit. The coplanar vectors p×L and (mk/r)r are shown in blue and green, respectively; these variables are defined below. The vector A is constant in direction and magnitude.*

$$\mathrm{A} = \mathrm{p} \times \mathrm{L} - mk\hat{\mathrm{r}}$$

where

- m is the mass of the point particle moving under the central force,
- P is its momentum vector,
- $\mathrm{L} = \mathrm{r} \times \mathrm{p}$ is its angular momentum vector,
- k is a parameter that describes strength of the central force,
- r is the position vector of the particle, and
- $\hat{\mathrm{r}}$ is the corresponding unit vector, i.e., $\hat{\mathrm{r}} = \dfrac{\mathrm{r}}{r}$ where r is the magnitude of r.

Since the assumed force is conservative, the total energy E is a constant of motion

$$E = \frac{p^2}{2m} - \frac{k}{r} = \frac{1}{2}mv^2 - \frac{k}{r}$$

Furthermore, the assumed force is a central force, and thus the angular momentum vector *L* is also conserved and defines the plane in which the particle travels. The LRL vector A is perpendicular to the angular momentum vector L because both $p \times L$ and r are perpendicular to *L*. It follows that A lies in the plane of the orbit.

This definition of the *LRL* vector A pertains to a single point particle of mass m moving under the action of a fixed force. However, the same definition may be extended to two-body problems such as Kepler's problem, by taking m as the reduced mass of the two bodies and r as the vector between the two bodies.

A variety of alternative formulations for the same constant of motion may also be used. The most common is to scale by mk to define the eccentricity vector

$$\mathrm{e} = \frac{\mathrm{A}}{mk} = \frac{1}{mk}(\mathrm{p} \times \mathrm{L}) - \hat{\mathrm{r}}$$

Derivation of the Kepler Orbits

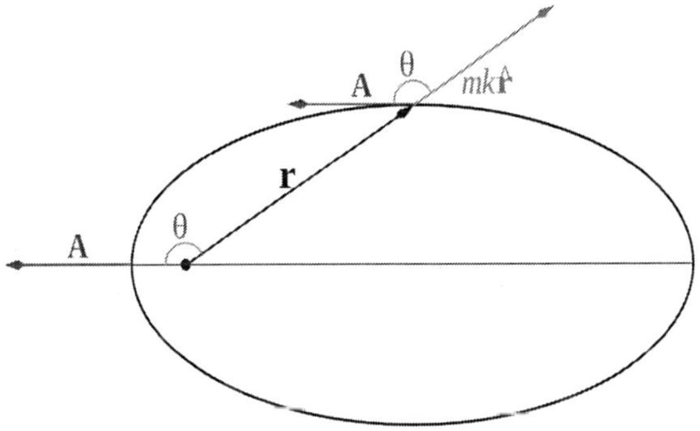

Figure: *Simplified version of Figure, defining the angle θ between A and r at one point of the orbit.*

The shape and orientation of the Kepler problem orbits can be determined from the LRL vector as follows. Taking the dot product of A with the position vector r gives the equation

$$\mathrm{A} \cdot \mathrm{r} = Ar\cos\theta = \mathrm{r} \cdot (\mathrm{p} \times \mathrm{L}) - mkr$$

where θ is the angle between r and A. Permuting the scalar triple product

$$\mathrm{r}\cdot(\mathrm{p}\times\mathrm{L})=(\mathrm{r}\times\mathrm{p})\cdot\mathrm{L}=\mathrm{L}\cdot\mathrm{L}=L^2$$

and rearranging yields the defining formula for a conic section

$$\frac{1}{r}=\frac{mk}{L^2}\left(1+\frac{A}{mk}\cos\theta\right)$$

of eccentricity e,

$$e=\frac{A}{mk}=\frac{|\mathrm{A}|}{mk}$$

and latus rectum

$$|4p|=\frac{2L^2}{mk}$$

The major semiaxis a of the conic section may be defined using the latus rectum and the eccentricity

$$a\left(1\pm e^2\right)=2p=\frac{L^2}{mk}$$

where the minus sign pertains to ellipses and the plus sign to hyperbolae.

Taking the dot product of A with itself yields an equation involving the energy E

$$A^2=m^2k^2+2mEL^2$$

which may be rewritten in terms of the eccentricity

$$e^2-1=\frac{2L^2}{mk^2}E$$

Thus, if the energy E is negative (bound orbits), the eccentricity is less than one and the orbit is an ellipse. Conversely, if the energy is positive (unbound orbits, also called "scattered orbits"), the eccentricity is greater than one and the orbit is a hyperbola.

Finally, if the energy is exactly zero, the eccentricity is one and the orbit is a parabola. In all cases, the direction of A lies along the symmetry axis of the conic section and points from the centre of force towards the periapsis, the point of closest approach.

Circular Momentum Hodographs

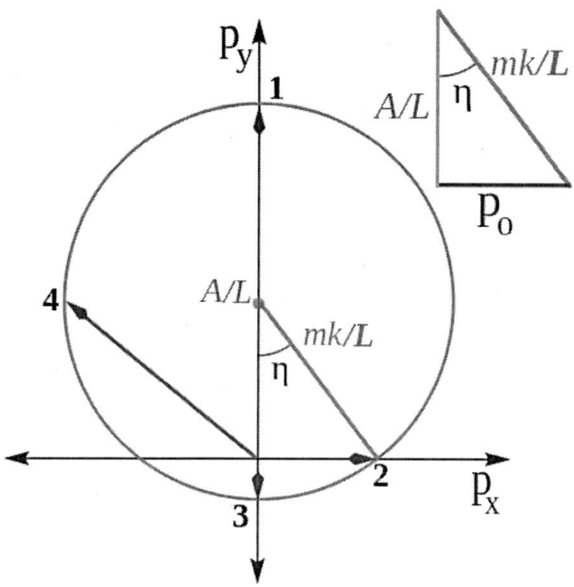

Figure: *The momentum vector p (shown in blue) moves on a circle as the particle moves on an ellipse. The four labelled points correspond to those in Figure. The circle is centred on the y-axis at position A/L (shown in magenta), with radius mk/L (shown in green). The angle η determines the eccentricity e of the elliptical orbit ($\cos\eta = e$). By the inscribed angle theorem for circles, η is also the angle between any point on the circle and the two points of intersection with the p_x axis, $p_x = \pm p_0$.*

The conservation of the *LRL* vector *A* and angular momentum vector L is useful in showing that the momentum vector p moves on a circle under an inverse-square central force.

Taking the dot product of

$$mk\hat{\mathrm{r}} = \mathrm{p} \times \mathrm{L} - \mathrm{A}$$

with itself yields

$$(mk)^2 = A^2 + p^2L^2 + 2\mathrm{L}\cdot(\mathrm{p} \times \mathrm{A}) .$$

Further choosing L along the z-axis, and the major semiaxis as the x-axis, yields the locus equation for p,

$$p_x^2 + \left(p_y - A / L\right)^2 = \left(mk / L\right)^2 .$$

In other words, the momentum vector p is confined to a circle of radius mk/L centred on (0, A/L). The eccentricity e corresponds to the cosine of the angle η. In the degenerate limit of circular orbits, and thus vanishing A, the circle centres at the origin (0,0). For brevity, it is also useful to introduce the variable $p_0 = \sqrt{2m|E|}$. This circular hodograph is useful in illustrating the symmetry of the Kepler problem.

Constants of Motion and Superintegrability

The seven scalar quantities E, A and L (being vectors, the latter two contribute three conserved quantities each) are related by two equations, $A \cdot L = 0$ and $A^2 = m^2k^2 + 2\, m\, E\, L^2$, giving five independent constants of motion. This is consistent with the six initial conditions (the particle's initial position and velocity vectors, each with three components) that specify the orbit of the particle, since the initial time is not determined by a constant of motion. Since the magnitude of A (and the eccentricity e of the orbit) can be determined from the total angular momentum L and the energy E, only the direction of A is conserved independently; moreover, since A must be perpendicular to L, it contributes only one additional conserved quantity.

A mechanical system with d degrees of freedom can have at most 2d – 1 constants of motion, since there are $2d$ initial conditions and the initial time cannot be determined by a constant of motion. A system with more than d constants of motion is called superintegrable and a system with $2d - 1$ constants is called maximally superintegrable. Since the solution of the Hamilton–Jacobi equation in one coordinate system can yield only d constants of motion, superintegrable systems must be separable in more than one coordinate system. The Kepler problem is maximally superintegrable, since it has three degrees of freedom (d=3) and five independent constant of motion; its Hamilton–Jacobi equation is separable in both spherical coordinates and parabolic coordinates, as described below.

Maximally superintegrable systems follow closed, one-dimensional orbits in phase space, since the orbit is the intersection of the phase-space isosurfaces of their constants of motion. Consequently, the orbits are perpendicular to all gradients of all these independent isosurfaces, five in this specific problem, and hence are determined by the generalized cross products of all of these gradients. As a result, all superintegrable systems are automatically describable by Nambu mechanics, alternatively, and equivalently, to Hamiltonian mechanics. Maximally superintegrable systems can be quantized using commutation relations, as illustrated below. Nevertheless, equivalently, they are also quantized in the Nambu framework, such as this classical Kepler problem into the quantum hydrogen atom.

Evolution Under Perturbed Potentials

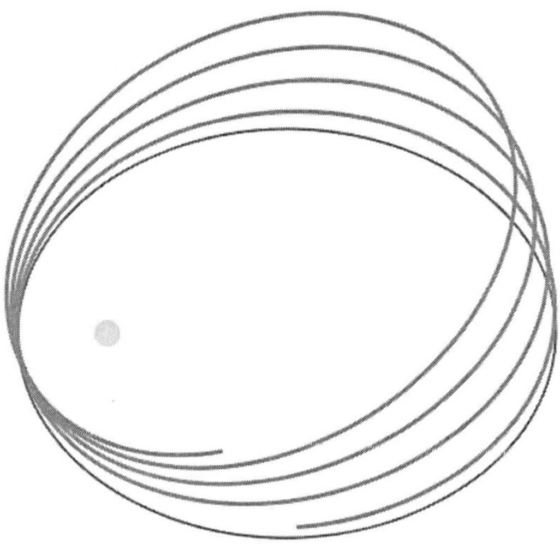

Figure: *Gradually precessing elliptical orbit, with an eccentricity e = 0.667. Such precession arises in the Kepler problem if the attractive central force deviates slightly from an inverse-square law. The rate of precession can be calculated using the formulae in the text.*

The Laplace–Runge–Lenz vector A is conserved only for a perfect inverse-square central force. In most practical problems such as planetary motion, however, the interaction potential energy between two bodies is not exactly an inverse square law, but may include an additional central force, a so-called perturbation described by a potential energy h(r). In such cases, the LRL vector rotates slowly in the plane of the orbit, corresponding to a slow apsidal precession of the orbit. By assumption, the perturbing potential h(r) is a conservative central force, which implies that the total energy E and angular momentum vector L are conserved. Thus, the motion still lies in a plane perpendicular to L and the magnitude A is conserved, from the equation $A^2 = m^2k^2 + 2mEL^2$. The perturbation potential h(r) may be any sort of function, but should be significantly weaker than the main inverse-square force between the two bodies.

The rate at which the LRL vector rotates gives information about the perturbing potential h(r). Using canonical perturbation theory and action-angle coordinates, it is straightforward to show that A rotates at a rate of

$$\begin{aligned}\frac{\partial}{\partial L}\langle h(r)\rangle &= \frac{\partial}{\partial L}\left\{\frac{1}{T}\int_0^T h(r)dt\right\} \\ &= \frac{\partial}{\partial L}\left\{\frac{m}{L^2}\int_0^{2\pi} r^2 h(r)d\theta\right\}.\end{aligned}$$

where T is the orbital period and the identity $L\,dt = m\,r^2\,d\theta$ was used to convert the time integral into an angular integral. The expression in angular brackets, 0h(r)
0, represents the perturbing potential, but averaged over one full period; that is, averaged over one full passage of the body around its orbit. Mathematically, that time average corresponds to the following quantity in curly braces. This averaging helps to suppress fluctuations in the rate of rotation.

This approach was used to help verify Einstein's theory of general relativity, which adds a small inverse-cubic perturbation to the normal Newtonian gravitational potential.

$$h(r) = \frac{kL^2}{m^2c^2}\left(\frac{1}{r^3}\right)$$

Inserting this function into the integral and using the equation

$$\frac{1}{r} = \frac{mk}{L^2}\left(1 + \frac{A}{mk}\cos\theta\right)$$

to express r in terms of θ, the precession rate of the periapsis caused by this non-Newtonian perturbation is calculated to be

$$\frac{6\pi k^2}{TL^2c^2}$$

which closely matches the observed anomalous precession of Mercury and binary pulsars. This agreement with experiment is considered to be strong evidence for general relativity.

Poisson Brackets

The three components L_i of the angular momentum vector L have the Poisson brackets

$$\{L_i, L_j\} = \sum_{s=1}^{3} \epsilon_{ijs} L_s \ ,$$

where i=1,2,3 and ε_{ijs} is the fully antisymmetric tensor, i.e., the Levi-Civita symbol; the summation index s is used here to avoid confusion with the force parameter k defined above. The Poisson brackets are represented here as square brackets (not curly braces), both for consistency with the references and because they will be interpreted as quantum mechanical commutation relations in the next section and as Lie brackets in a following section.

As noted below, a scaled Laplace–Runge–Lenz vector D may be defined with the same units as angular momentum by dividing A by $p_0 = \sqrt{2m\,|\,E\,|}$. The Poisson brackets of D with the angular momentum vector L can then be written in a similar form

$$\{D_i, L_j\} = \sum_{s=1}^{3} \epsilon_{ijs} D_s \ .$$

The Poisson brackets of *D* with itself depend on the sign of E, i.e., on whether the total energy *E* is negative (producing closed, elliptical orbits under an inverse-square central force) or positive (producing open, hyperbolic orbits under an inverse-square central force). For negative energies – i.e., for bound systems – the Poisson brackets are

$$\{D_i, D_j\} = \sum_{s=1}^{3} \epsilon_{ijs} L_s \ ,$$

whereas, for positive energy, the Poisson brackets have the opposite sign,.

$$\{D_i, D_j\} = -\sum_{s=1}^{3} \epsilon_{ijs} L_s \ .$$

The Casimir invariants for negative energies are

$$C_1 = \mathrm{D{\cdot}D} + \mathrm{L{\cdot}L} = \frac{mk^2}{2|E|}$$

$$C_2 = \mathrm{D{\cdot}L} = 0,$$

and have vanishing Poisson brackets with all components of D and L,

$$\{C_1, L_i\} = \{C_1, D_i\} = \{C_2, L_i\} = \{C_2, D_i\} = 0 \ .$$

C_2 is trivially zero, since the two vectors are always perpendicular.

However, the other invariant, C_1, is non-trivial and depends only on m, k and E. Upon canonical quantization, this invariant allows the energy levels of hydrogen-like atoms to be derived using only quantum mechanical canonical commutation relations, instead of the conventional solution of the Schrödinger equation.

Quantum Mechanics of the Hydrogen Atom

Poisson brackets provide a simple guide for quantizing most classical systems: the commutation relation of two quantum mechanical operators is specified by the Poisson bracket of the corresponding classical variables, multiplied by i'.

By carrying out this quantization and calculating the eigenvalues of the C_1 Casimir operator for the Kepler problem, Wolfgang Pauli was able to derive the energy levels of hydrogen-like atoms and, thus, their atomic emission spectrum.

This elegant derivation was obtained before the development of the Schrödinger equation.

A subtlety of the quantum mechanical operator for the LRL vector A is that the momentum and angular momentum operators do not commute; hence, the cross product of p and L must be defined carefully. Typically, the operators for the Cartesian components A_s are defined using a symmetrized product,

$$A_s = -mk\hat{r}_s + \frac{1}{2}\sum_{i=1}^{3}\sum_{j=1}^{3}\epsilon_{sij}\left(p_i l_j + l_j p_i\right),$$

from which the corresponding additional ladder operators for L can be defined,

$$J_0 = A_3$$

$$J_{\pm 1} = \mp\frac{1}{\sqrt{2}}\left(A_1 \pm iA_2\right).$$

These further connect different eigenstates of L^2.

A normalized first Casimir invariant operator, quantum analog of the above, can likewise be defined,

$$C_1 = -\frac{mk^2}{2\hbar^2}H^{-1} - I,$$

where H^{-1} is the inverse of the Hamiltonian energy operator, and I is the identity operator.

Applying these ladder operators to the eigenstates $|lmn\rangle$ of the total angular momentum, azimuthal angular momentum and energy operators, the eigenvalues of the first Casimir operator C_1 are $n^2 - 1$. Importantly, by dint of the vanishing of C_2, they are independent of the l and m quantum numbers, making the energy levels degenerate.

Hence, the energy levels are given by

$$E_n = -\frac{mk^2}{2\hbar^2 n^2},$$

which coincides with the Rydberg formula for hydrogen-like atoms. The additional symmetry operators A have implicitly connected the different l multiplets among themselves, for a given energy (and C_1). In effect, they have enlarged the group SO(3) to SO(4).

Conservation and Symmetry

The conservation of the LRL vector corresponds to a subtle symmetry of the system. In classical mechanics, symmetries are continuous operations that map one orbit onto another without changing the energy of the system; in quantum mechanics, symmetries are continuous operations that "mix" electronic orbitals of the same energy, i.e., degenerate energy levels. A conserved quantity is usually associated with such symmetries. For example, every central force is symmetric under the rotation group SO(3), leading to the conservation of angular momentum L. Classically, an overall rotation of the system does not affect the energy of an orbit; quantum mechanically, rotations mix the spherical harmonics of the same quantum number l without changing the energy.

The symmetry for the inverse-square central force is higher and more subtle. The peculiar symmetry of the Kepler problem results in the conservation of both the angular momentum vector L and the LRL vector A (as defined above) and, quantum mechanically, ensures that the energy levels of hydrogen do not depend on the angular momentum quantum numbers l and m. The symmetry is more subtle, however, because the symmetry operation must take place in a higher-dimensional space; such symmetries are often called "hidden symmetries". Classically, the higher symmetry of the Kepler problem allows for continuous alterations of the orbits that preserve energy but not angular momentum; expressed another way, orbits of the same energy but different angular momentum (eccentricity) can be transformed continuously into one another. Quantum mechanically, this corresponds to mixing orbitals that differ in the l and m quantum numbers, such as the s (l=0) and p (l=1) atomic orbitals. Such mixing cannot be done with ordinary three-dimensional translations or rotations, but is equivalent to a rotation in a higher dimension.

For negative energies – i.e., for bound systems – the higher symmetry group is SO(4), which preserves the length of four-dimensional vectors

$$|\mathrm{e}|^2 = e_1^2 + e_2^2 + e_3^2 + e_4^2$$

In 1935, Vladimir Fock showed that the quantum mechanical bound Kepler problem is equivalent to the problem of a free particle confined to a three-dimensional unit sphere in four-dimensional space. Specifically, Fock showed that the Schrödinger wavefunction in the momentum space for the Kepler problem was the stereographic projection of the spherical harmonics on the sphere. Rotation of the sphere and reprojection results in a continuous mapping of the elliptical orbits without changing the energy; quantum mechanically, this corresponds to a mixing of all orbitals of the same energy quantum number n. Valentine Bargmann noted subsequently that the Poisson brackets for the angular momentum vector L and the scaled LRL vector D formed the Lie algebra for SO(4). Simply put, the six quantities D and L correspond to the six conserved angular momenta in four dimensions, associated with the six possible simple rotations in that space (there are six ways of choosing two axes from four).

This conclusion does not imply that our universe is a three-dimensional sphere; it merely means that this particular physics problem (the two-body problem for inverse-square central forces) is mathematically equivalent to a free particle on a three-dimensional sphere.

For positive energies – i.e., for unbound, "scattered" systems – the higher symmetry group is $SO(3,1)$, which preserves the Minkowski length of 4-vectors

$$ds^2 = e_1^2 + e_2^2 + e_3^2 - e_4^2$$

Both the negative- and positive-energy cases were considered by Fock and Bargmann and have been reviewed encyclopedically by Bander and Itzykson.

The orbits of central-force systems – and those of the Kepler problem in particular – are also symmetric under reflection. Therefore, the $SO(3)$, $SO(4)$ and $SO(3,1)$ groups cited above are not the full symmetry groups of their orbits; the full groups are O(3), O(4) and O(3,1), respectively. Nevertheless, only the connected subgroups, $SO(3)$, $SO(4)$ and $SO(3,1)$, are needed to demonstrate the conservation of the angular momentum and LRL vectors; the reflection symmetry is irrelevant for conservation, which may be derived from the Lie algebra of the group.

Rotational Symmetry in four Dimensions

The connection between the Kepler problem and four-dimensional rotational symmetry SO(4) can be readily visualized. Let the four-dimensional Cartesian coordinates be denoted (w, x, y, z) where (x, y, z) represent the Cartesian coordinates of the normal position vector r. The three-dimensional momentum vector p is associated with a four-dimensional vector η on a three-dimensional unit sphere

$$\begin{aligned}\eta &= \frac{p^2 - p_0^2}{p^2 + p_0^2}\hat{\mathrm{w}} + \frac{2p_0}{p^2 + p_0^2}\mathrm{p} \\ &= \frac{mk - rp_0^2}{mk}\hat{\mathrm{w}} + \frac{rp_0}{mk}\mathrm{p}\end{aligned}$$

where $\hat{\mathrm{w}}$ is the unit vector along the new w-axis. The transformation mapping p to η can be uniquely inverted; for example, the x-component of the momentum equals

$$p_x = p_0 \frac{\eta_x}{1 - \eta_w}$$

and similarly for p_y and p_z. In other words, the three-dimensional vector p is a stereographic projection of the four-dimensional η vector, scaled by p_0.

Without loss of generality, we may eliminate the normal rotational symmetry by choosing the Cartesian coordinates such that the z-axis is aligned with the angular momentum vector L and the momentum hodographs are aligned as they are in Figure, with the centres of the circles on the y-axis. Since the motion is planar, and p and L are perpendicular, $p_z = \eta_z = 0$ and attention may be restricted to the three-dimensional vector $\eta = (\eta_w, \eta_x, \eta_y)$. The family of Apollonian circles of momentum hodographs correspond to a family of great circles on the three-dimensional η sphere, all of which intersect the η_x-axis at the two foci $\eta_x = \pm 1$, corresponding to the momentum hodograph foci at $p_x = \pm p_0$. These great circles are related by a simple rotation about the η_x-axis. This rotational symmetry transforms all the orbits of the same energy into one another; however, such a rotation is orthogonal to the usual three-dimensional rotations, since it transforms the fourth dimension η_w. This higher symmetry is characteristic of the Kepler problem and corresponds to the conservation of the LRL vector.

An elegant action-angle variables solution for the Kepler problem can be obtained by eliminating the redundant four-dimensional coordinates in favour of elliptic cylindrical coordinates (χ, ψ, φ)

$$\eta_w = \text{cn}\chi\ \text{cn}\psi$$

$$\eta_x = \text{sn}\chi\ \text{dn}\psi\ \cos\phi$$

$$\eta_y = \text{sn}\chi\ \text{dn}\psi\ \sin\phi$$

$$\eta_z = \text{dn}\chi\ \text{sn}\psi$$

where sn, cn and dn are Jacobi's elliptic functions.

Generalizations to Other Potentials and Relativity

The Laplace–Runge–Lenz vector can also be generalized to identify conserved quantities that apply to other situations.

In the presence of an electric field E, the conserved generalized Laplace–Runge–Lenz vector $\mathcal{A}$ is

$$\mathcal{A} = \mathrm{A} + \frac{mq}{2}\left[(\mathrm{r}\times\mathrm{E})\times\mathrm{r}\right]$$

where q is the charge on the orbiting particle.

Further generalizing the Laplace–Runge–Lenz vector to other potentials and special relativity, the most general form can be written as

$$\mathcal{A} = \left(\frac{\partial\xi}{\partial u}\right)(\mathrm{p}\times\mathrm{L}) + \left[\xi - u\left(\frac{\partial\xi}{\partial u}\right)\right]L^2\hat{\mathrm{r}}$$

where u = 1/r (cf. Bertrand's theorem) and $\xi = \cos\theta$, with the angle θ defined by

$$\theta = L\int^u \frac{du}{\sqrt{m^2c^2\left(\gamma^2-1\right)-L^2u^2}}$$

and γ is the Lorentz factor. As before, we may obtain a conserved binormal vector $\mathcal{B}$ by taking the cross product with the conserved angular momentum vector

$$\mathcal{B} = \mathrm{L}\times\mathcal{A}$$

These two vectors may likewise be combined into a conserved dyadic tensor W

$$\mathcal{W} = \alpha\mathcal{A}\otimes\mathcal{A} + \beta\mathcal{B}\otimes\mathcal{B}$$

For illustration, the LRL vector for a non-relativistic, isotropic harmonic oscillator can be calculated. Since the force is central

$$F(r) = -kr$$

the angular momentum vector is conserved and the motion lies in a plane. The conserved dyadic tensor can be written in a simple form

$$\mathcal{W} = \frac{1}{2m}\mathrm{p}\otimes\mathrm{p} + \frac{k}{2}\mathrm{r}\otimes\mathrm{r}$$

although it should be noted that p and r are not necessarily perpendicular. The corresponding Runge–Lenz vector is more complicated

$$\mathcal{A} = \frac{1}{\sqrt{mr^2\omega_0 A - mr^2 E + L^2}}\{(\mathrm{p}\times\mathrm{L}) + (mr\omega_0 A - mrE)\hat{\mathrm{r}}\}$$

where $\omega_0 = \sqrt{\frac{k}{m}}$ is the natural oscillation frequency and $A = (E^2 - \omega^2 L^2)^{1/2} / \omega$.

Proofs that the Laplace–Runge–Lenz vector is Conserved in Kepler Problems

The following are arguments showing that the LRL vector is conserved under central forces that obey an inverse-square law.

Direct Proof of Conservation

A central force Facting on the particle is

$$\mathrm{F} = \frac{d\mathrm{p}}{dt} = f(r)\frac{\mathrm{r}}{r} = f(r)\hat{\mathrm{r}}$$

for some function $f(r)$ of the radius r. Since the angular momentum $\mathrm{L} = \mathrm{r}\times\mathrm{p}$ is conserved under central forces,

$$\frac{d}{dt}\mathrm{L} = 0$$

and

$$\frac{d}{dt}(\mathrm{p}\times\mathrm{L}) = \frac{d\mathrm{p}}{dt}\times\mathrm{L} = f(r)\hat{\mathrm{r}}\times\left(\mathrm{r}\times m\frac{d\mathrm{r}}{dt}\right) = f(r)\frac{m}{r}\left[\mathrm{r}\left(\mathrm{r}\cdot\frac{d\mathrm{r}}{dt}\right) - r^2\frac{d\mathrm{r}}{dt}\right]$$

where the momentum $\mathrm{p} = m\frac{d\mathrm{r}}{dt}$ and where the triple cross product has been simplified using Lagrange's formula

$$\mathrm{r} \times \left(\mathrm{r} \times \frac{d\mathrm{r}}{dt} \right) = \mathrm{r} \left(\mathrm{r} \cdot \frac{d\mathrm{r}}{dt} \right) - r^2 \frac{d\mathrm{r}}{dt}$$

The identity

$$\frac{d}{dt}(\mathrm{r} \cdot \mathrm{r}) = 2\mathrm{r} \cdot \frac{d\mathrm{r}}{dt} = \frac{d}{dt}\left(r^2\right) = 2r\frac{dr}{dt}$$

yields the equation

$$\frac{d}{dt}(\mathrm{p} \times \mathrm{L}) = -mf(r)r^2 \left[\frac{1}{r}\frac{d\mathrm{r}}{dt} - \frac{\mathrm{r}}{r^2}\frac{dr}{dt} \right] = -mf(r)r^2 \frac{d}{dt}\left(\frac{\mathrm{r}}{r} \right)$$

For the special case of an inverse-square central force $f(r) = \frac{-k}{r^2}$, this equals

$$\frac{d}{dt}(\mathrm{p} \times \mathrm{L}) = mk\frac{d}{dt}\left(\frac{\mathrm{r}}{r} \right) = \frac{d}{dt}(mk\hat{\mathrm{r}})$$

Therefore, A is conserved for inverse-square central forces

$$\frac{d}{dt}\mathrm{A} = \frac{d}{dt}(\mathrm{p} \times \mathrm{L}) - \frac{d}{dt}(mk\hat{\mathrm{r}}) = 0$$

A shorter proof is obtained by using the relation of angular momentum to angular velocity, $\mathrm{L} = mr^2\omega$, which holds for a particle travelling in a plane perpendicular to *L*. Specifying to inverse-square central forces, the time derivative of $\mathrm{p} \times \mathrm{L}$ is

$$\frac{d}{dt}\mathrm{p} \times \mathrm{L} = \left(\frac{-k}{r^2}\hat{\mathrm{r}} \right) \times \left(mr^2\omega \right) = mk\omega \times \hat{\mathrm{r}} = mk\frac{d}{dt}\hat{\mathrm{r}}$$

where the last equality holds because a unit vector can only change by rotation, and $\omega \times \hat{\mathrm{r}}$ is the orbital velocity of the rotating vector.

Thus, A is seen to be a difference of two vectors with equal time derivatives.

As described below, this LRL vector A is a special case of a general conserved vector *A*that can be defined for all central forces.

However, since most central forces do not produce closed orbits, the analogous vector *A*rarely has a simple definition and is generally a multivalued function of the angle θ between *r* and *A*.

Hamilton–Jacobi Equation in Parabolic Coordinates

The constancy of the LRL vector can also be derived from the Hamilton–Jacobi equation in parabolic coordinates (ξ, η), which are defined by the equations

$$\xi = r + x$$

$$\eta = r - x$$

where r represents the radius in the plane of the orbit

$$r = \sqrt{x^2 + y^2}$$

The inversion of these coordinates is

$$x = \frac{1}{2}(\xi - \eta)$$

$$y = \sqrt{\xi\eta}$$

Separation of the Hamilton–Jacobi equation in these coordinates yields the two equivalent equations

$$2\xi p_\xi^2 - mk - mE\xi = -\Gamma$$

$$2\eta p_\eta^2 - mk - mE\eta = \Gamma$$

where Γ is a constant of motion. Subtraction and re-expression in terms of the Cartesian momenta p_x and p_y shows that Γ is equivalent to the LRL vector

$$\Gamma = p_y\left(xp_y - yp_x\right) - mk\frac{x}{r} = A_x$$

Noether's Theorem

The connection between the rotational symmetry described above and the conservation of the LRL vector can be made quantitative by way of Noether's theorem. This theorem, which is used for finding constants of motion, states that any infinitesimal variation of the generalized coordinates of a physical system

$$\delta q_i = \epsilon g_i(\mathrm{q}, \dot{\mathrm{q}}, t)$$

that causes the Lagrangian to vary to first order by a total time derivative

$$\delta L = \epsilon \frac{d}{dt} G(\mathrm{q}, t)$$

corresponds to a conserved quantity Γ

$$\Gamma = -G + \sum_i g_i \left(\frac{\partial L}{\partial \dot{q}_i} \right)$$

In particular, the conserved LRL vector component A_s corresponds to the variation in the coordinates

$$\delta x_i = \frac{\epsilon}{2}\left[2p_i x_s - x_i p_s - \delta_{is}\left(\mathbf{r}\cdot\mathbf{p}\right)\right]$$

where i equals 1, 2 and 3, with x_i and p_i being the ith components of the position and momentum vectors *r* and *p*, respectively; as usual, δ_{is} represents the Kronecker delta. The resulting first-order change in the Lagrangian is

$$\delta L = \epsilon mk \frac{d}{dt}\left(\frac{x_s}{r} \right)$$

Substitution into the general formula for the conserved quantity Γ yields the conserved component A_s of the LRL vector

$$A_s = \left[p^2 x_s - p_s\left(\mathbf{r}\cdot\mathbf{p}\right)\right] - mk\left(\frac{x_s}{r} \right) = \left[\mathbf{p}\times\left(\mathbf{r}\times\mathbf{p}\right)\right]_s - mk\left(\frac{x_s}{r} \right)$$

Lie Transformation

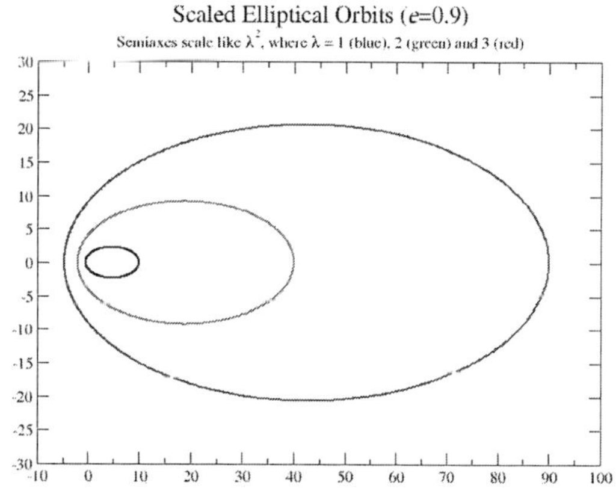

Figure: *The Lie transformation from which the conservation of the LRL vector A is derived. As the scaling parameter λ varies, the energy and angular momentum changes, but the eccentricity e and the magnitude and direction of A do not.*

The Noether theorem derivation of the conservation of the LRL vector A is elegant, but has one drawback: the coordinate variation δx_i involves not only the position *r*, but also the momentum p or, equivalently, the velocity *υ*. This drawback may be eliminated by instead deriving the conservation of A using an approach pioneered by Sophus Lie. Specifically, one may define a Lie transformation in which the coordinates r and the time t are scaled by different powers of a parameter λ.

$$t \to \lambda^3 t, \mathrm{r} \to \lambda^2 \mathrm{r}, \mathrm{p} \to \frac{1}{\lambda}\mathrm{p}$$

This transformation changes the total angular momentum *L* and energy *E*

$$L \to \lambda L, E \to \frac{1}{\lambda^2}E$$

but preserves their product EL^2. Therefore, the eccentricity e and the magnitude A are preserved, as may be seen from the equation for A^2

$$A^2 = m^2k^2e^2 = m^2k^2 + 2mEL^2$$

The direction of A is preserved as well, since the semiaxes are not altered by a global scaling. This transformation also preserves Kepler's third law, namely, that the semiaxis a and the period T form a constant T^2/a^3.

Alternative Scalings, Symbols and Formulations

Unlike the momentum and angular momentum vectors p and L, there is no universally accepted definition of the Laplace–Runge–Lenz vector; several different scaling factors and symbols are used in the scientific literature. The most common definition is given above, but another common alternative is to divide by the constant mk to obtain a dimensionless conserved eccentricity vector

$$\mathrm{e} = \frac{1}{mk}(\mathrm{p} \times \mathrm{L}) - \hat{\mathrm{r}} = \frac{m}{k}(\mathrm{v} \times (\mathrm{r} \times \mathrm{v})) - \hat{\mathrm{r}}$$

where v is the velocity vector. This scaled vector e has the same direction as *A* and its magnitude equals the eccentricity of the orbit. Other scaled versions are also possible, e.g., by dividing *A* by m alone

$$\mathrm{M} = \mathrm{v} \times \mathrm{L} - k\hat{\mathrm{r}}$$

or by p_0

$$\mathrm{D} = \frac{\mathrm{A}}{p_0} = \frac{1}{\sqrt{2m|E|}}\{\mathrm{p} \times \mathrm{L} - mk\hat{\mathrm{r}}\}$$

which has the same units as the angular momentum vector L. In rare cases, the sign of the LRL vector may be reversed, i.e., scaled by −1. Other common symbols for the LRL vector include a, R, F, J and V. However, the choice of scaling and symbol for the LRL vector do not affect its conservation.

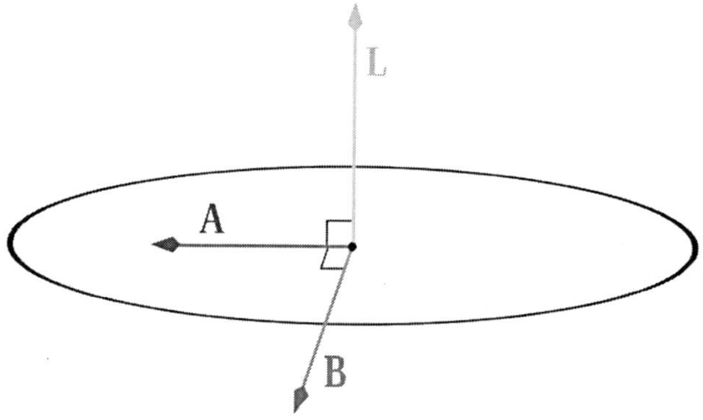

Figure: *The angular momentum vector L, the LRL vector A and Hamilton's vector, the binormal B, are mutually perpendicular; A and B point along the major and minor axes, respectively, of an elliptical orbit of the Kepler problem.*

An alternative conserved vector is the binormal vector B studied by William Rowan Hamilton

$$\mathrm{B} = \mathrm{p} - \left(\frac{mk}{L^2 r}\right)(\mathrm{L} \times \mathrm{r})$$

which is conserved and points along the minor semiaxis of the ellipse; the LRL vector $A = B \times \mathrm{L}$ is the cross product of B and L. The vector B is denoted as "binormal" since it is perpendicular to both A and L. Similar to the LRL vector itself, the binormal vector can be defined with different scalings and symbols.

The two conserved vectors, A and B can be combined to form a conserved dyadic tensor W

$$W = \alpha A \otimes A + \beta B \otimes B$$

where α and β are arbitrary scaling constants and $\otimes$ represents the tensor product (which is not related to the vector cross product, despite their similar symbol). Written in explicit components, this equation reads

$$W_{ij} = \alpha A_i A_j + \beta B_i B_j$$

Being perpendicular to each another, the vectors A and B can be viewed as the principal axes of the conserved tensor W, i.e., its scaled eigenvectors. W is perpendicular to L

$$L \cdot W = \alpha(L \cdot A)A + \beta(L \cdot B)B = 0$$

since A and B are both perpendicular to L as well, $L \circ A = L \cdot B = 0$. For clarification, this equation reads in explicit components

$$(L \cdot W)_j = \alpha\left(\sum_{i=1}^{3} L_i A_i\right)A_j + \beta\left(\sum_{i=1}^{3} L_i B_i\right)B_j = 0$$

Quadruple Product

In mathematics, the quadruple product is a product of four vectors in three-dimensional Euclidean space. The name "quadruple product" is used for two different products, the scalar-valued scalar quadruple product and the vector-valued vector quadruple product.

Scalar Quadruple Product

The scalar quadruple product is defined as the dot product of two cross products:

$$(a \times b) \cdot (c \times d),$$

where a, b, c, d are vectors in three-dimensional Euclidean space. It can be evaluated using the identity:

$$(a \times b) \cdot (c \times d) = (a \cdot c)(b \cdot d) - (a \cdot d)(b \cdot c).$$

or using the determinant:

$$(a \times b) \cdot (c \times d) = \begin{vmatrix} a \cdot c & a \cdot d \\ b \cdot c & b \cdot d \end{vmatrix}.$$

Vector Quadruple Product

The vector quadruple product is defined as the cross product of two cross products:

$$(\mathrm{a} \times \mathrm{b}) \times (\mathrm{c} \times \mathrm{d}),$$

where a, b, c, d are vectors in three-dimensional Euclidean space. It can be evaluated using the identity:

$$(\mathrm{a} \times \mathrm{b}) \times (\mathrm{c} \times \mathrm{d}) = [\mathrm{a, b, d}]\mathrm{c} - [\mathrm{a, b, c}]\mathrm{d},$$

This identity can also be written using tensor notation and the Einstein summation convention as follows:

$$(\mathrm{a} \times \mathrm{b}) \times (\mathrm{c} \times \mathrm{d}) = \varepsilon_{ijk} a^i c^j d^k b^l - \varepsilon_{ijk} b^i c^j d^k a^l = \varepsilon_{ijk} a^i b^j d^k c^l - \varepsilon_{ijk} a^i b^j c^k d^l$$

using the notation for the triple product:

$$[\mathrm{a, b, d}] = (\mathrm{a} \times \mathrm{b}) \cdot \mathrm{d} = \begin{vmatrix} \mathrm{a}\cdot\hat{\mathrm{i}} & \mathrm{b}\cdot\hat{\mathrm{i}} & \mathrm{d}\cdot\hat{\mathrm{i}} \\ \mathrm{a}\cdot\hat{\mathrm{j}} & \mathrm{b}\cdot\hat{\mathrm{j}} & \mathrm{d}\cdot\hat{\mathrm{j}} \\ \mathrm{a}\cdot\hat{\mathrm{k}} & \mathrm{b}\cdot\hat{\mathrm{k}} & \mathrm{d}\cdot\hat{\mathrm{k}} \end{vmatrix} = \begin{vmatrix} \mathrm{a}\cdot\hat{\mathrm{i}} & \mathrm{a}\cdot\hat{\mathrm{j}} & \mathrm{a}\cdot\hat{\mathrm{k}} \\ \mathrm{b}\cdot\hat{\mathrm{i}} & \mathrm{b}\cdot\hat{\mathrm{j}} & \mathrm{b}\cdot\hat{\mathrm{k}} \\ \mathrm{d}\cdot\hat{\mathrm{i}} & \mathrm{d}\cdot\hat{\mathrm{j}} & \mathrm{d}\cdot\hat{\mathrm{k}} \end{vmatrix},$$

where the last two forms are determinants with $\hat{\mathrm{i}}, \hat{\mathrm{j}}, \hat{\mathrm{k}}$ denoting unit vectors along three mutually orthogonal directions.

Equivalent forms can be obtained using the identity:

$$[\mathrm{b, c, d}]\mathrm{a} - [\mathrm{c, d, a}]\mathrm{b} + [\mathrm{d, a, b}]\mathrm{c} - [\mathrm{a, b, c}]\mathrm{d} = 0.$$

Application

The quadruple products are useful for deriving various formulas in spherical and plane geometry. For example, if four points are chosen on the unit sphere, *A, B, C, D*, and unit vectors drawn from the centre of the sphere to the four points, *a, b, c,* d respectively, the identity:

$$(\mathrm{a} \times \mathrm{b}) \cdot (\mathrm{c} \times \mathrm{d}) - (\mathrm{a}\cdot\mathrm{c})(\mathrm{b}\cdot\mathrm{d}) \quad (\mathrm{a}\cdot\mathrm{d})(\mathrm{b}\cdot\mathrm{c}),$$

in conjunction with the relation for the magnitude of the cross product:

$$\| \mathrm{a} \times \mathrm{b} \| = ab \sin\theta_{ab},$$

and the dot product:

$$\| \mathrm{a}\cdot\mathrm{b} \| = ab \cos\theta_{ab},$$

where a = b = 1 for the unit sphere, results in the identity among the angles attributed to Gauss:

$$\sin\theta_{ab}\sin\theta_{cd}\cos x = \cos\theta_{ac}\cos\theta_{bd} - \cos\theta_{ad}\cos\theta_{bc},$$

where x is the angle between a × b and c × d, or equivalently, between the planes defined by these vectors.

Josiah Willard Gibbs's pioneering work on vector calculus provides several other examples.

Right-hand Rule

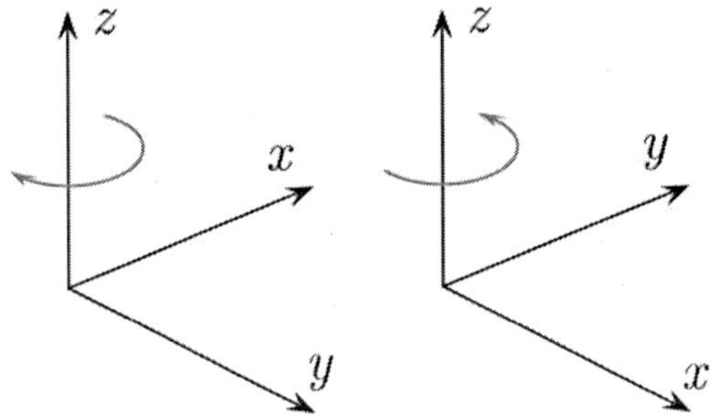

***Figure:** The left-handed orientation is shown on the left, and the right-handed on the right.*

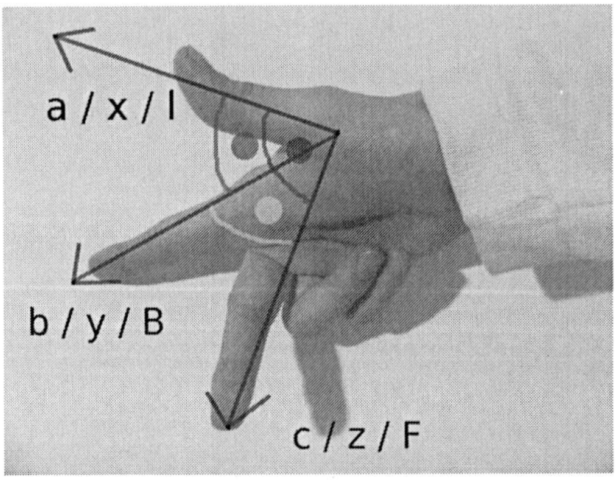

***Figure:** Use of right hand*

In mathematics and physics, the right-hand rule is a common mnemonic for understanding notation conventions for vectors in 3 dimensions. It was invented for use in electromagnetism by British physicist John Ambrose Fleming in the late 19th century. When choosing three vectors that must be at right angles to each other, there are two distinct solutions, so when expressing this idea in mathematics, one must remove the ambiguity of which solution is meant. There are variations on the mnemonic depending on context, but all variations are related to the one idea of choosing a convention.

The right-handed Screw

Curl the fingers of your right hand around a screw and hold it with your thumb pointing up. Turn the screw counter-clockwise, as viewed looking down upon the tip of your extended thumb. The direction of this turning motion can also be described as turning in the direction from the base of your fingers towards the tips of your fingers. If it moves in the direction of your thumb, which is up in this case, the screw is called right-handed. If it moves into your hand, which is the opposite direction that your thumb is pointing, the screw would be left-handed. Most screws are right-handed.

Right-handed and Left-handed Coordinates

Let the X and Y axes define a horizontal plane with the X axis pointing towards you. Then the Z axis can either point up (right-handed) or down (left-handed). If the coordinates are right-handed and you place your right fist on the plane then your fingers will curl from the first or X axis to the second or Y axis and your thumb will point along the Z axis. If the coordinates were left handed the fingers of your left hand would curl from the first axis to the second and your left thumb would point along the Z axis.

If any one axis is reversed the handedness changes. To preserve handedness one of the other axes must also reverse which is equivalent to switching the labels on the other two axes. Note that the handedness of coordinates has no meaning unless the axes are labelled in sequence: 1,2,3 or x,y,z.

Direction Associated with an Ordered Pair of Directions

One form of the right-hand rule is used in situations in which an ordered operation must be performed on two vectors a and b that has a result which is a vector c perpendicular to both a and b. The most common example is the vector cross product. The right-hand rule imposes the following procedure for choosing one of the two directions. $\vec{a} \times \vec{b} = \vec{c}$

- With the thumb, index, and middle fingers at right angles to each other (with the index finger pointed straight), the middle finger points in the direction of c when the thumb represents a and the index finger represents b.

Other (equivalent) finger assignments are possible. For example, the first (index) finger can represent a, the first vector in the product; the second (middle) finger, b, the second vector; and the thumb, c, the product.

Direction Associated with a Rotation

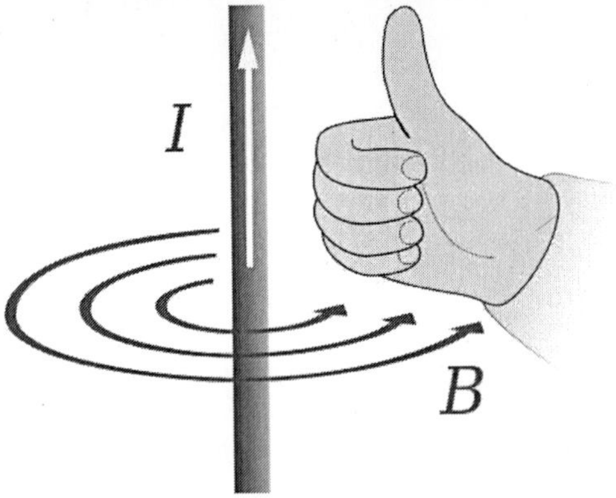

Figure: *Prediction of direction of field (B), given that the current I flows in the direction of the thumb*

Figure: *The right-hand rule as applied to motion produced with screw threads*

A different form of the right-hand rule, sometimes called the right-hand grip rule or the corkscrew-rule or the right-hand thumb rule, is used in situations where a vector must be assigned to the rotation of a body, a magnetic field or a fluid.

Alternatively, when a rotation is specified by a vector, and it is necessary to understand the way in which the rotation occurs, the right-hand grip rule is applicable.

This version of the rule is used in two complementary applications of Ampère's circuital law:

1. An electric current passes through a solenoid, resulting in a magnetic field. When you wrap your right hand around the solenoid with your fingers in the direction of the conventional current, your thumb points in the direction of the magnetic north pole.
2. An electric current passes through a straight wire. Here, the thumb points in the direction of the conventional current (from positive to negative), and the fingers point in the direction of the magnetic lines of flux.

The principle is also used to determine the direction of the torque vector. If you grip the imaginary axis of rotation of the rotational force so that your fingers point in the direction of the force, then the extended thumb points in the direction of the torque vector.

The right-hand grip rule is a convention derived from the right-hand rule convention for vectors. When applying the rule to current in a straight wire for example, the direction of the magnetic field (counterclockwise instead of clockwise when viewed from the tip of the thumb) is a result of this convention and not an underlying physical phenomenon.

Applications

The first form of the rule is used to determine the direction of the cross product of two vectors. This leads to widespread use in physics, wherever the cross product occurs. A list of physical quantities whose directions are related by the right-hand rule is given below. (Some of these are related only indirectly to cross products, and use the second form.)

- The angular velocity of a rotating object and the rotational velocity of any point on the object
- A torque, the force that causes it, and the position of the point of application of the force
- A magnetic field, the position of the point where it is determined, and the electric current (or change in electric flux) that causes it
- A magnetic field in a coil of wire and the electric current in the wire
- The force of a magnetic field on a charged particle, the magnetic field itself, and the velocity of the object
- The vorticity at any point in the field of flow of a fluid
- The induced current from motion in a magnetic field (known as Fleming's right-hand rule)
- The x, y and z unit vectors in a Cartesian coordinate system can be chosen to follow the right-hand rule. Right-handed coordinate systems are often used in rigid body physics and kinematics.

Fleming's left-hand rule is a rule for finding the direction of the thrust on a conductor carrying a current in a magnetic field.

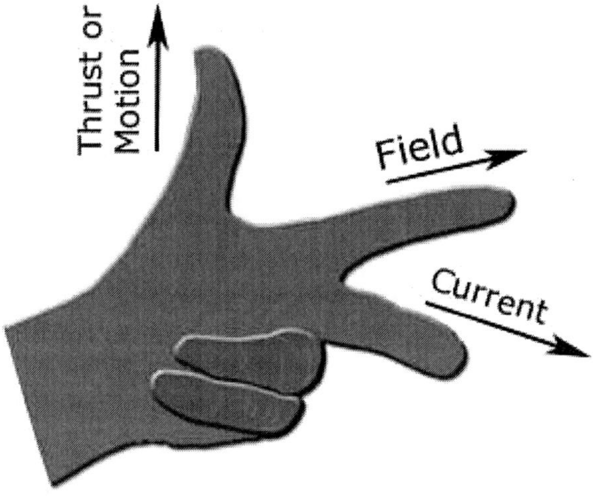

Figure: *Fleming's left-hand rule*

Left-hand Rule

In certain situations, it may be useful to use the opposite convention, where one of the vectors is reversed and so creates a left-handed triad instead of a right-handed triad.

An example of this situation is for left-handed materials. Normally, for an electromagnetic wave, the electric and magnetic fields, and the direction of propagation of the wave obey the right-hand rule. However, left-handed materials have special properties, notably the negative refractive index. It makes the direction of propagation point in the opposite direction.

De Graaf's translation of Fleming's left-hand rule - which uses thrust, field and current - and the right-hand rule, is the FBI rule. The FBI rule changes thrust into F (Lorentz force), B (direction of the magnetic field) and I (current). The FBI rule is easily remembered by US citizens because of the commonly known abbreviation for the Federal Bureau of Investigation.

Symmetry

Vector	*Right-hand*	*Right-hand*	*Right-hand*	*Left-hand*	*Left-hand*	*Left-Hand*
a, x or I	Thumb	Fingers or palm	First or Index	Thumb	Fingers or palm	First or index
b, y or B	First or index	Thumb	Fingers or palm	Fingers or palm	First or index	Thumb
c, z or F	Fingers or palm	First or index	Thumb	First or index	Thumb	Fingers or palm

Chapter 3

Row Vector

In linear algebra, a row vector or row matrix is a 1 × m matrix, i.e. a matrix consisting of a single row of m elements:

$$\mathrm{x} = \begin{bmatrix} x_1 & x_2 & \dots & x_m \end{bmatrix}.$$

The transpose of a row vector is a column vector:

$$\begin{bmatrix} x_1 \; x_2 \dots x_m \end{bmatrix}^{\mathrm{T}} = \begin{bmatrix} x_1 \\ x_2 \\ \vdots \\ x_m \end{bmatrix}.$$

The set of all row vectors forms a vector space which acts like the dual space to the set of all column vectors, in the sense that any linear functional on the space of column vectors (i.e. any element of the dual space) can be represented uniquely as a dot product with a specific row vector.

Notation

Row vectors are sometimes written using the following non-standard notation:

$$\mathrm{x} = \begin{bmatrix} x_1, x_2, \dots, x_m \end{bmatrix}.$$

Operations

- Matrix multiplication involves the action of multiplying each row vector of one matrix by each column vector of another matrix.
- The dot product of two vectors a and b is equivalent to multiplying the row vector representation of a by the column vector representation of b:

$$a{\cdot}b = \begin{bmatrix} a_1 & a_2 & a_3 \end{bmatrix} \begin{bmatrix} b_1 \\ b_2 \\ b_3 \end{bmatrix}.$$

Preferred Input Vectors for Matrix Transformations

Frequently a row vector presents itself for an operation within n-space expressed by an n by n matrix M:

$$v\ M = p.$$

Then p is also a row vector and may present to another n by n matrix Q:

$$p\ Q = t.$$

Conveniently, one can write $t = p\ Q = v\ MQ$ telling us that the matrix product transformation MQ can take v directly to t. Continuing with row vectors, matrix transformations further reconfiguring n-space can be applied to the right of previous outputs. In contrast, when a column vector is transformed to become another column under an n by n matrix action, the operation occurs to the left:

$$p = M\ v \text{ and } t = Q\ p\ ,$$

leading to the algebraic expression $QM\ v$ for the composed output from v input. The matrix transformations mount up to the left in this use of a column vector for input to matrix transformation. The natural bias to read left-to-right, as subsequent transformations are applied in linear algebra, stands against column vector inputs.

Nevertheless, using the transpose operation these differences between inputs of a row or column nature are resolved by an antihomomorphism between the groups arising on the two sides. The technical construction uses the dual space associated with a vector space to develop the transpose of a linear map. For an instance where this row vector input convention has been used to good effect see Raiz Usmani, where on page 106 the convention allows the statement "The product mapping ST of U into W [is given] by:

$$\alpha(ST) = (\alpha S)T = \beta T = \gamma\text{."}$$

(The Greek letters represent row vectors).

Ludwik Silberstein used row vectors for spacetime events; he applied Lorentz transformation matrices on the right in his

Theory of Relativity in 1914. In 1963 when McGraw-Hill published Differential Geometry by Heinrich Guggenheimer of the University of Minnesota, he uses the row vector convention in chapter 5, "Introduction to transformation groups" (eqs. 7a,9b and 12 to 15). When H. S. M. Coxeter reviewed Linear Geometry by Rafael Artzy, he wrote, "[Artzy] is to be congratulated on his choice of the 'left-to-right' convention, which enables him to regard a point as a row matrix instead of the clumsy column that many authors prefer."

Scalar Projection

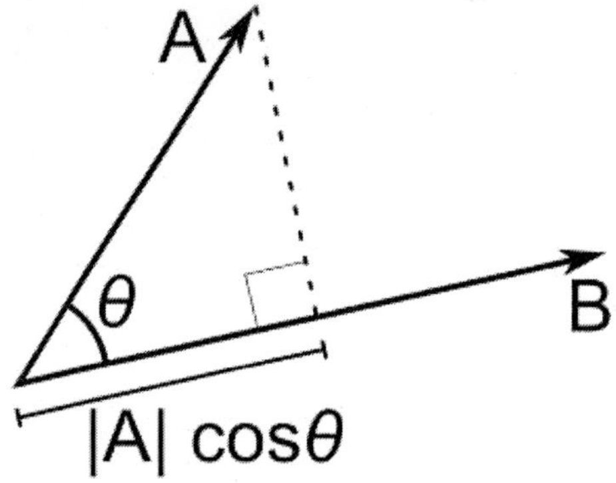

Figure: *If 0° ≤ θ ≤ 90°, as in this case, the scalar projection of a on b coincides with the length of the vector projection.*

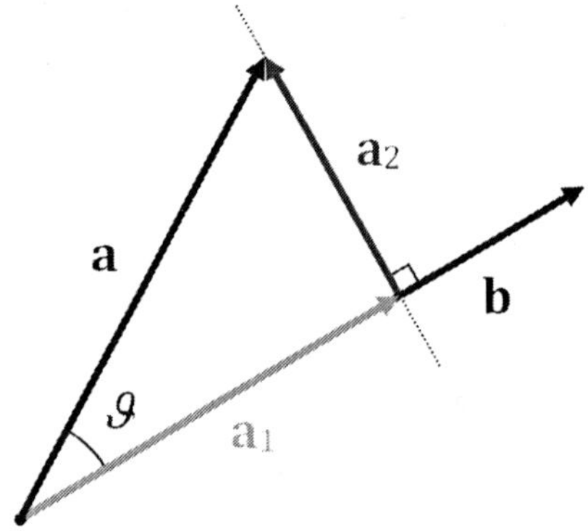

Figure: *Vector projection of a on b (a_1), and vector rejection of a from b (a_2).*

In mathematics, the scalar projection of a vector a on (or onto) a vector b, also known as the scalar resolute or scalar component of in the direction of b, is given by:

$$s = |a| \cos\theta = a \cdot \hat{b},$$

where the operator denotes a dot product, $\hat{b}$ is the unit vector in the direction of b, $|a|$ is the length of a, and $\theta$is the angle between a and b.

The scalar projection is a scalar, equal to the length of the orthogonal projection of a on b, with a minus sign if the projection has an opposite direction with respect to b.

Multiplying the scalar projection of a on b by $\hat{b}$ converts it into the above-mentioned orthogonal projection, also called vector projection of *a* on *b*.

Definition Based on Angle θ

If the angle between and is known, the scalar projection of on can be computed using

Definition in Terms of a and b

When $\theta$is not known, the cosine of θcan be computed in terms of a and b, by the following property of the dot product $a \cdot b$:

$$\frac{a \cdot b}{|a||b|} = \cos\theta$$

By this property, the definition of the scalar projection becomes:

$$s = |a| \cos\theta = |a| \frac{a \cdot b}{|a||b|} = \frac{a \cdot b}{|b|}$$

Properties

The scalar projection has a negative sign if $90 < \theta < 180$ degrees. It coincides with the length of the corresponding vector projection if the angle is smaller than 90°. More exactly, if the vector projection is denoted a_1 and its length $|a_1|$:

$s = |a_1|$ if $0 < \theta \leq 90$ degrees,

$s = -|a_1|$ if $90 < \theta \leq 180$ degrees.

Seven-dimensional Cross Product

In mathematics, the seven-dimensional cross product is a bilinear operation on vectors in seven dimensional Euclidean space. It assigns to any two vectors a, b in $\mathbb{R}^7$ a vector $a \times b$ also in $\mathbb{R}^7$. Like the cross product in three dimensions the seven-dimensional product is anticommutative and $a \times b$ is orthogonal to both a and b. Unlike in three dimensions, it does not satisfy the Jacobi identity. And while the three-dimensional cross product is unique up to a change in sign, there are many seven-dimensional cross products. The seven-dimensional cross product has the same relationship to octonions as the three-dimensional product does to quaternions.

The seven-dimensional cross product is one way of generalising the cross product to other than three dimensions, and it turns out to be the only other non-trivial bilinear product of two vectors that is vector valued, anticommutative and orthogonal. In other dimensions there are vector-valued products of three or more vectors that satisfy these conditions, and binary products with bivector results.

Definition

We can define a cross product on a Euclidean space V as a bilinear map from $V \times V$ to V mapping vectors x and y in V to another vector x × y also in V, where $x \times y$ has the properties

- orthogonality:

$$x \cdot (x \times y) = (x \times y) \cdot y = 0,$$

- magnitude:

$$|x \times y|^2 = |x|^2 |y|^2 - (x \cdot y)^2$$

where $(x \cdot y)$ is the Euclidean dot product and $|x|$ is the vector norm. The first property states that the cross product is perpendicular to its arguments, while the second property gives the magnitude of the cross product. An equivalent expression in terms of the angle θ between the vectors is

$$|x \times y| = |x|\,|y| \sin\theta,$$

or the area of the parallelogram in the plane of x and y with the two vectors as sides. As a third alternative the following can be shown to be equivalent to either expression for the magnitude:

$$|x \times y| = |x|\,|y| \text{ if } (x \cdot y) = 0.$$

Consequences of the Defining Properties

Given the three basic properties of (i) bilinearity, (ii) orthogonality and (iii) magnitude discussed in the section on definition, a nontrivial cross product exists only in three and seven dimensions. This restriction upon dimensionality can be shown by postulating the properties required for the cross product, then deducing a equation which is only satisfied when the dimension is 0, 1, 3 or 7. In zero dimensions there is only the zero vector, while in one dimension all vectors are parallel, so in both these cases a cross product must be identically zero.

The restriction to 0, 1, 3 and 7 dimensions is related to Hurwitz's theorem, that normed division algebras are only possible in 1, 2, 4 and 8 dimensions. The cross product is derived from the product of the algebra by considering the product restricted to the 0, 1, 3, or 7 imaginary dimensions of the algebra. Again discarding trivial products the product can only be defined this way in three and seven dimensions.

In contrast with three dimensions where the cross product is unique (apart from sign), there are many possible binary cross products in seven dimensions. One way to see this is to note that given any pair of vectors x and $y \in \mathbb{R}^7$ and any vector v of magnitude $|v| = |x|\,|y| \sin\theta$ in the five dimensional space perpendicular to the plane spanned by x and y, it is possible to find a cross product with a multiplication table (and an associated set of basis vectors) such that $x \times y = v$. That leaves open the question of just how many vector pairs like x and y can be matched to specified directions like v before the limitations of any particular table intervene.

Another difference between the three dimensional cross product and a seven dimensional cross product is:

"...for the cross product $x \times y$ in $\mathbb{R}^7$ there are also other planes than the linear span of x and y giving the same direction as $x \times y$" —Pertti Lounesto, Clifford algebras and spinors, p. 97 This statement is exemplified by every multiplication table, because any specific unit vector selected as a product occurs as a mapping from three different pairs of unit vectors, once with a plus sign and once with a minus sign. Each of these different pairs, of course, corresponds to another plane being mapped into the same direction.

Further properties follow from the definition, including the following identities:

1. Anticommutativity: $x \times y = -y \times x$,
2. Scalar triple product: $x \cdot (y \times z) = y \cdot (z \times x) = z \cdot (x \times y)$
3. Malcev identity:

$$(x \times y) \times (x \times z) = ((x \times y) \times z) \times x + ((y \times z) \times x) \times x + ((z \times x) \times x) \times y$$

$$x \times (x \times y) = -|x|^2 y + (x \cdot y)x.$$

Other properties follow only in the three dimensional case, and are not satisfied by the seven dimensional cross product, notably,

1. Vector triple product:

$$x \times (y \times z) = (x \cdot z)y - (x \cdot y)z$$

2. Jacobi identity:

$$x \times (y \times z) + y \times (z \times x) + z \times (x \times y) = 0$$

Different Multiplication Tables

Two different multiplication tables have been used in this article, and there are more. These multiplication tables are characterized by the Fano plane, and these are shown in the figure for the two tables used here: at top, the one described by Sabinin, Sbitneva, and Shestakov, and at bottom that described by Lounesto. The numbers under the Fano diagrams (the set of lines in the diagram) indicate a set of indices for seven independent products in each case, interpreted as ijk → $e_i \times e_j = e_k$. The multiplication table is recovered from the Fano diagram by following either the straight line connecting any three points, or the circle in the centre, with a sign as given by the arrows. For example, the first row of multiplications resulting in e_1 in the above listing is obtained by following the three paths connected to e_1 in the lower Fano diagram: the circular path $e_2 \times e_4$, the diagonal path $e_3 \times e_7$, and the edge path $e_6 \times e_1 = e_5$ rearranged using one of the above identities as:

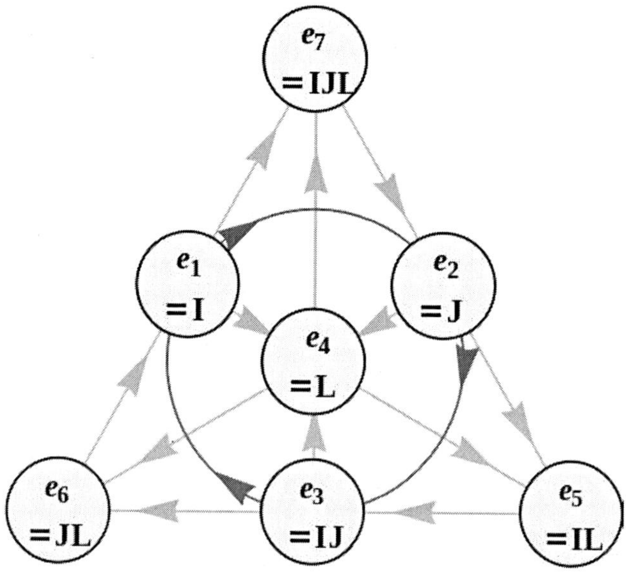

123,145,176,246,257,347,365

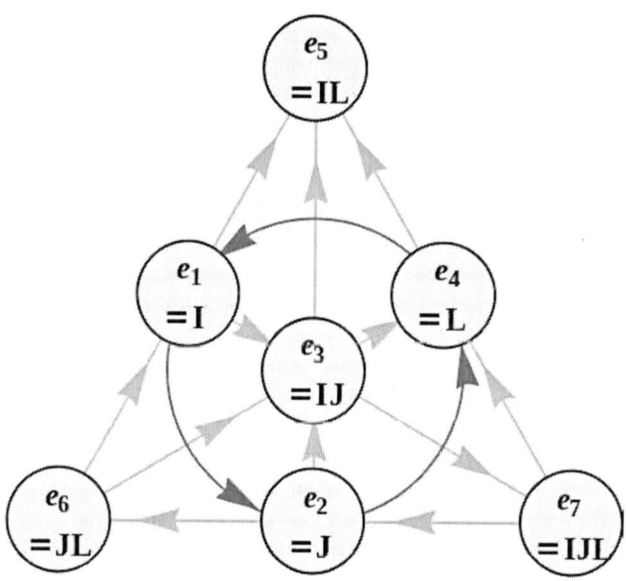

124,137,156,235,267,346,457

Figure: *Fano planes for the two multiplication tables used here.*

$$e_6 \times (e_6 \times e_1) = -e_1 = e_6 \times e_5,$$

or

$$e_5 \times e_6 = e_1,$$

also obtained directly from the diagram with the rule that any two unit vectors on a straight line are connected by multiplication to the third unit vector on that straight line with signs according to the arrows (sign of the permutation that orders the unit vectors).

It can be seen that both multiplication rules follow from the same Fano diagram by simply renaming the unit vectors, and changing the sense of the centre unit vector. The question arises: how many multiplication tables are there?

The question of possible multiplication tables arises, for example, when one reads another article on octonions, which uses a different one from the one given by [Cayley, say]. Usually it is remarked that all 480 possible ones are equivalent, that is, given an octonionic algebra with a multiplication table and any other valid multiplication table, one can choose a basis such that the multiplication follows the new table in this basis. One may also take the point of view, that there exist different octonionic algebras, that is, algebras with different multiplication tables. With this interpretation...all these octonionic algebras are isomorphic.

—Jörg Schray, Corinne A Manogue, Octonionic representations of Clifford algebras and triality (1994)

Using Geometric Algebra

The product can also be calculated using geometric algebra. The product starts with the exterior product, a bivector valued product of two vectors:

$$B = x \wedge y = \frac{1}{2}(xy - yx).$$

This is bilinear, alternate, has the desired magnitude, but is not vector valued. The vector, and so the cross product, comes from the product of this bivector with a trivector. In three dimensions up to a scale factor there is only one trivector, the pseudoscalar of the space, and a product of the above bivector and one of the two unit trivectors gives the vector result, the dual of the bivector. A similar calculation is done is seven dimensions, except as trivectors form a 35-dimensional space there are many trivectors that could be used, though not just any trivector will do. The trivector that gives the same product as the above coordinate transform is

$$v = e_{124} + e_{235} + e_{346} + e_{457} + e_{561} + e_{672} + e_{713}.$$

This is combined with the exterior product to give the cross product

$$x \times y = -(x \wedge y) \lrcorner v$$

where $\lrcorner$ is the left contraction operator from geometric algebra.

Relation to the Octonions

Just as the 3-dimensional cross product can be expressed in terms of the quaternions, the 7-dimensional cross product can be expressed in terms of the octonions. After identifying $\mathbb{R}^7$ with the imaginary octonions (the orthogonal complement of the real line in O), the cross product is given in terms of octonion multiplication by

$$x \times y = \mathrm{Im}(xy) = \frac{1}{2}(xy - yx).$$

Conversely, suppose V is a 7-dimensional Euclidean space with a given cross product. Then one can define a bilinear multiplication on $\mathbb{R} \oplus V$ as follows:

$$(a, x)(b, y) = (ab - x \cdot y, ay + bx + x \times y).$$

The space $\mathbb{R} \oplus V$ with this multiplication is then isomorphic to the octonions.

The cross product only exists in three and seven dimensions as one can always define a multiplication on a space of one higher dimension as above, and this space can be shown to be a normed division algebra. By Hurwitz's theorem such algebras only exist in one, two, four, and eight dimensions, so the cross product must be in zero, one, three or seven dimensions. The products in zero and one dimensions are trivial, so non-trivial cross products only exist in three and seven dimensions. The failure of the 7-dimension cross product to satisfy the Jacobi identity is due to the nonassociativity of the octonions. In fact,

$$x \times (y \times z) + y \times (z \times x) + z \times (x \times y) = -\frac{3}{2}[x, y, z]$$

where [x, y, z] is the associator.

Rotations

In three dimensions the cross product is invariant under the group of the rotation group, SO(3), so the cross product of x and y after they are rotated is the image of x × y under the rotation. But this invariance is not true in seven dimensions; that is, the cross product is not invariant under the group of rotations in seven dimensions, SO(7). Instead it is invariant under the exceptional Lie group G_2, a subgroup of SO(7).

Generalizations

Non-trivial binary cross products exist only in three and seven dimensions. But if the restriction that the product is binary is lifted, so products of more than two vectors are allowed, then more products are possible. As in two dimensions the product must be vector valued, linear, and anti-commutative in any two of the vectors in the product.

The product should satisfy orthogonality, so it is orthogonal to all its members. This means no more than n – 1 vectors can be used in n dimensions. The magnitude of the product should equal the volume of the parallelotope with the vectors as edges, which is can be calculated using the Gram determinant. So the conditions are

- orthogonality:

$$(a_1 \times \cdots \times a_k) \cdot a_j = 0$$

- the Gram determinant:

$$|a_1 \times \cdots \times a_k|^2 = \det(a_i \cdot a_j) = \begin{vmatrix} a_1 \cdot a_1 & a_1 \cdot a_2 & \dots & a_1 \cdot a_k \\ a_2 \cdot a_1 & a_2 \cdot a_2 & \dots & a_2 \cdot a_k \\ \dots & \dots & \dots & \dots \\ a_k \cdot a_1 & a_k \cdot a_2 & \dots & a_k \cdot a_k \end{vmatrix}$$

The Gram determinant is the squared volume of the parallelotope with a_1, ..., a_k as edges. If there are just two vectors x and y it simplifies to the condition for the binary cross product given above, that is

$$|\mathbf{x}\times\mathbf{y}|^2=\begin{vmatrix}\mathbf{x}\cdot\mathbf{x} & \mathbf{x}\cdot\mathbf{y}\\ \mathbf{y}\cdot\mathbf{x} & \mathbf{y}\cdot\mathbf{y}\end{vmatrix}=|\mathbf{x}|^2|\mathbf{y}|^2-(\mathbf{x}\cdot\mathbf{y})^2,$$

With these conditions a non-trivial cross product only exists:

- as a binary product in three and seven dimensions
- as a product of $n - 1$ vectors in $n > 3$ dimensions
- as a product of three vectors in eight dimensions

The product of n – 1 vectors is in n dimensions is the Hodge dual of the exterior product of n – 1 vectors. One version of the product of three vectors in eight dimensions is given by

$$a \times b \times c = (a \wedge b \wedge c) \lrcorner\, (w - ve_8)$$

where v is the same trivector as used in seven dimensions, is again the left contraction, and $w = \lrcorner\, ve_{12...7}$ is a 4-vector.

Vector Notation

This page is an overview of the common notations used when working with vectors, which may be spatial or more abstract members of vector spaces.

The common typographic convention for representing a vector is upright boldface type, as in v for a vector named 'v'. In handwriting, where boldface type is either unavailable or unwieldy, vectors are often represented with right-pointing arrows or harpoons above their names, as in $\vec{v}$. Shorthand notations include tildes and straight lines placed above or below the name of a vector.

Rectangular Vectors

A rectangular vector is a coordinate vector specified by components that define a rectangle (or rectangular prism in three dimensions, and similar shapes in greater dimensions). The starting point and terminal point of the vector lie at opposite ends of the rectangle (or prism, etc.).

Ordered Set Notation: A rectangular vector in $\mathbb{R}^n$ can be specified using an ordered set of components, enclosed in either parentheses or angle brackets.

In a general sense, an n-dimensional vector v can be specified in either of the following forms:

- $\mathrm{v} = (v_1, v_2, \ldots, v_{n-1}, v_n)$
- $\mathrm{v} = \langle v_1, v_2, \ldots, v_{n-1}, v_n \rangle$

Where v_1, v_2, ..., v_{n-1}, v_n are the components of v.

Matrix Notation: A rectangular vector in $\mathbb{R}^n$ can also be specified as a row or column matrix containing the ordered set of components. A vector specified as a row matrix is known as a row vector; one specified as a column matrix is known as a column vector.

Again, an n-dimensional vector $v$can be specified in either of the following forms using matrices:

- $\mathrm{v} = \begin{bmatrix} v_1 & v_2 & \cdots & v_{n-1} & v_n \end{bmatrix} = \begin{pmatrix} v_1 & v_2 & \cdots & v_{n-1} & v_n \end{pmatrix}$
- $\mathrm{v} = \begin{bmatrix} v_1 \\ v_2 \\ \vdots \\ v_{n-1} \\ v_n \end{bmatrix} = \begin{pmatrix} v_1 \\ v_2 \\ \vdots \\ v_{n-1} \\ v_n \end{pmatrix}$

Where v_1, v_2, ..., $\mathrm{v}_{n"1}$, v_n are the components of v. In some advanced contexts, a row and a column vector have different meaning.

Engineering Notation: A rectangular vector in $\mathbb{R}^3$ (or fewer dimensions, such as $\mathbb{R}^2$ where v_z below is zero) can be specified as the sum of the scalar multiples of the components of the vector with the members of the standard basis in $\mathbb{R}^3$. The basis is represented with the unit vectors $\hat{\imath} = (1,0,0)$, $\hat{\jmath} = (0,1,0)$, and $\hat{\mathrm{k}} = (0,0,1)$.

A three-dimensional vector v can be specified in the following form, using engineering notation:

- $\mathrm{v} = v_x \hat{\imath} + v_y \hat{\jmath} + v_z \hat{\mathrm{k}}$

Where v_x, v_y, and v_z are the magnitudes of the components of v.

Polar Vectors

A polar vector is a vector in two dimensions specified as a magnitude (or length) and a direction (or angle). It is akin to an arrow in the polar coordinate system. The magnitude, typically represented as r, is the length from the starting point of the vector to its endpoint. The angle, typically represented as θ (the Greek letter theta), is measured as the offset from the horizontal (or a line collinear with the x-axis in the positive direction). The angle is typically reduced to lie within the range $0 \le \theta < 2\pi$ radians or $0 \le \theta < 360^\circ$.

Ordered Set and Matrix Notations: Polar vectors can be specified using either ordered pair notation (a subset of ordered set notation using only two components) or matrix notation, as with rectangular vectors. In these forms, the first component of the vector is r (instead of v_1) and the second component is θ (instead of v_2). To differentiate polar vectors from rectangular vectors, the angle may be prefixed with the angle symbol, $\angle$.

A two-dimensional polar vector v can be represented as any of the following, using either ordered pair or matrix notation:

- $\mathrm{v} = (r, \angle\theta)$
- $\mathrm{v} = \langle r, \angle\theta \rangle$
- $\mathrm{v} = \begin{bmatrix} r & \angle\theta \end{bmatrix}$
- $\mathrm{v} = \begin{bmatrix} r \\ \angle\theta \end{bmatrix}$

Where r is the magnitude, θ is the angle, and the angle symbol ($\angle$) is optional.

Direct Notation: Polar vectors can also be specified using simplified autonomous equations that define r and θ explicitly. This can be unwieldy, but is useful for avoiding the confusion with two-dimensional rectangular vectors that arises from using ordered pair or matrix notation.

A two-dimensional vector whose magnitude is 5 units and whose direction is π/9 radians (20°) can be specified using either of the following forms:

- $r = 5, \theta = \dfrac{\pi}{9}$
- $r = 5, \theta = 20^\circ$

Cylindrical Vectors

A cylindrical vector is an extension of the concept of polar vectors into three dimensions. It is akin to an arrow in the cylindrical coordinate system. A cylindrical vector is specified by a distance in the xy-plane, an angle, and a distance from the xy-plane (a height). The first distance, usually represented as r or ρ (the Greek letter rho), is the magnitude of the projection of the vector onto the xy-plane. The angle, usually represented as θ or φ (the Greek letter phi), is measured as the offset from the line collinear with the x-axis in the positive direction; the angle is typically reduced to lie within the range $0 \leq \theta < 2\pi$. The second distance, usually represented as h or z, is the distance from the xy-plane to the endpoint of the vector.

Ordered set and Matrix Notations: Cylindrical vectors are specified like polar vectors, where the second distance component is concatenated as a third component to form ordered triplets (again, a subset of ordered set notation) and matrices. The angle may be prefixed with the angle symbol ($\angle$); the distance-angle-distance combination distinguishes cylindrical vectors in this notation from spherical vectors in similar notation.

A three-dimensional cylindrical vector v can be represented as any of the following, using either ordered triplet or matrix notation:

- $\mathrm{v} = (r, \angle\theta, h)$
- $\mathrm{v} = \langle r, \angle\theta, h \rangle$
- $\mathrm{v} = \begin{bmatrix} r & \angle\theta & h \end{bmatrix}$
- $\mathrm{v} = \begin{bmatrix} r \\ \angle\theta \\ h \end{bmatrix}$

Where r is the magnitude of the projection of v onto the xy-plane, θ is the angle between the positive x-axis and v, and h is the height from the xy-plane to the endpoint of v. Again, the angle symbol ($\angle$) is optional.

Direct Notation: A cylindrical vector can also be specified directly, using simplified autonomous equations that define r (or ρ), θ (or φ), and h (or z). Consistency should be used when choosing the names to use for the variables; ρ should not be mixed with θ and so on.

A three-dimensional vector, the magnitude of whose projection onto the xy-plane is 5 units, whose angle from the positive x-axis is $\pi/9$ radians (20°), and whose height from the xy-plane is 3 units can be specified in any of the following forms:

- $r = 5, \theta = \frac{\pi}{9}, h = 3$
- $r = 5, \theta = 20^{\circ}, h = 3$
- $\rho = 5, \phi = \frac{\pi}{9}, z = 3$
- $\rho = 5, \phi = 20^{\circ}, z = 3$

Spherical Vectors

A spherical vector is another method for extending the concept of polar vectors into three dimensions. It is akin to an arrow in the spherical coordinate system. A spherical vector is specified by a magnitude, an azimuth angle, and a zenith angle. The magnitude is usually represented as ρ. The azimuth angle, usually represented as θ, is the offset from the line collinear with the x-axis in the positive direction. The zenith angle, usually represented as φ, is the offset from the line collinear with the z-axis in the positive direction. Both angles are typically reduced to lie within the range from zero (inclusive) to 2π (exclusive).

Ordered Set and Matrix Notations: Spherical vectors are specified like polar vectors, where the zenith angle is concatenated as a third component to form ordered triplets and matrices. The azimuth and zenith angles may be both prefixed with the angle symbol ($\angle$); the prefix should be used consistently to produce the distance-angle-angle combination that distinguishes spherical vectors from cylindrical ones.

A three-dimensional spherical vector v can be represented as any of the following, using either ordered triplet or matrix notation:

- $\mathrm{v} = (\rho, \angle\theta, \angle\phi)$
- $\mathrm{v} = \langle\rho, \angle\theta, \angle\phi\rangle$
- $\mathrm{v} = \begin{bmatrix}\rho & \angle\theta & \angle\phi\end{bmatrix}$

- $v = \begin{bmatrix} \rho \\ \angle\theta \\ \angle\phi \end{bmatrix}$

Where ρ is the magnitude, θ is the azimuth angle, and φ is the zenith angle.

Direct Notation: Like polar and cylindrical vectors, spherical vectors can be specified using simplified autonomous equations, in this case for ρ, θ, and φ.

A three-dimensional vector whose magnitude is 5 units, whose azimuth angle is π/9 radians (20°), and whose zenith angle is π/4 radians (45°) can be specified as:

- $\rho = 5, \theta = \frac{\pi}{9}, \phi = \frac{\pi}{4}$
- $\rho = 5, \theta = 20^{\circ}, \phi = 45^{\circ}$

Operations

In any given vector space, the operations of vector addition and scalar multiplication are defined. Normed vector spaces also define an operation known as the norm (or determination of magnitude). Inner product spaces also define an operation known as the inner product. In $\mathbb{R}^n$, the inner product is known as the dot product. In $\mathbb{R}^3$ and $\mathbb{R}^7$, an additional operation known as the cross product is also defined.

Vector Addition

Vector addition is represented with the plus sign used as an operator between two vectors. The sum of two vectors u and v would be represented as:

$$u + v$$

Scalar Multiplication

Scalar multiplication is represented in the same manners as algebraic multiplication. A scalar beside a vector (either or both of which may be in parentheses) implies scalar multiplication. The two common operators, a dot and a rotated cross, are also acceptable (although the rotated cross is almost never used), but they risk confusion with dot products and cross products, which operate on two vectors. The product of a scalar c with a vector v can be represented in any of the following fashions:

- $c\mathrm{v}$
- $c\cdot\mathrm{v}$
- $c\times\mathrm{v}$

Vector Subtraction and Scalar Division

Using the algebraic properties of subtraction and division, along with scalar multiplication, it is also possible to "subtract" two vectors and "divide" a vector by a scalar.

Vector subtraction is performed by adding the scalar multiple of –1 with the second vector operand to the first vector operand. This can be represented by the use of the minus sign as an operator. The difference between two vectors u and v can be represented in either of the following fashions:

- $\mathrm{u}+-\mathrm{v}$
- $\mathrm{u}-\mathrm{v}$

Scalar division is performed by multiplying the vector operand with the numeric inverse of the scalar operand. This can be represented by the use of the fraction bar or division signs as operators. The quotient of a vector v and a scalar c can be represented in any of the following forms:

- $\frac{1}{c}\mathrm{v}$
- $\frac{\mathrm{v}}{c}$
- $\mathrm{v}\div c$

Norm

The norm of a vector is represented with double bars on both sides of the vector. The norm of a vector v can be represented as:

$$\|\mathrm{v}\|$$

The norm is also sometimes represented with single bars, like $|\mathrm{v}|$, but this can be confused with absolute value (which is a type of norm).

Inner Product

The inner product (also known as the scalar product, not to be confused with scalar multiplication) of two vectors is represented as an ordered pair enclosed in angle brackets. The inner product of two vectors u and v would be represented as:

$$\langle u, v \rangle$$

Dot Product

In $\mathbb{R}^n$, the inner product is also known as the dot product. In addition to the standard inner product notation, the dot product notation (using the dot as an operator) can also be used (and is more common). The dot product of two vectors *u* and v can be represented as:

$$u \cdot v$$

In some older literature, the dot product is implied between two vectors written side-by-side. This notation can be confused with the dyadic product between two vectors.

Cross Product

The cross product of two vectors (in $\mathbb{R}^3$) is represented using the rotated cross as an operator. The cross product of two vectors *u* and *v* would be represented as:

$$u \times v$$

In some older literature, the following notation is used for the cross product between *u* and *v*:

$$[u, v]$$

Chapter 4

Vector Projection

The vector projection of a vector a on (or onto) a nonzero vector b (also known as the vector component or vector resolute of a in the direction of b) is the orthogonal projection of a onto a straight line parallel to *b*. It is a vector parallel to *b*, defined as

$$\mathrm{a}_1 = a_1 \hat{\mathrm{b}}$$

where a_1 is a scalar, called the scalar projection of a onto *b*, and *b* is the unit vector in the direction of *b*. In turn, the scalar projection is defined as

$$a_1 = |\mathrm{a}| \cos\theta = \mathrm{a} \cdot \hat{\mathrm{b}},$$

where the operator $\cdot$ denotes a dot product, $|\mathrm{a}|$ is the length of a, and θ is the angle between a and b. The scalar projection is equal to the length of the vector projection, with a minus sign if the direction of the projection is opposite to the direction of *b*.

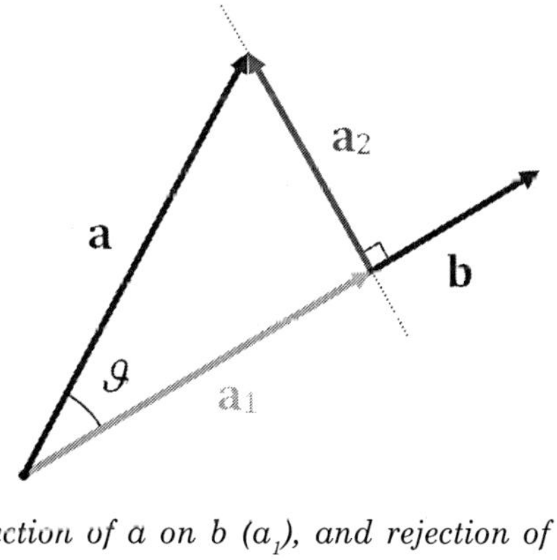

Figure: *Projection of a on b (a_1), and rejection of a from b (a_2).*

The vector component or vector resolute of a perpendicular to b, sometimes also called the vector rejection of a from b, is the orthogonal projection of a onto the plane (or, in general, hyperplane) orthogonal to b. Both the projection a_1 and rejection a_2 of a vector a are vectors, and their sum is equal to a, which implies that the rejection is given by

$$a_2 = a - a_1.$$

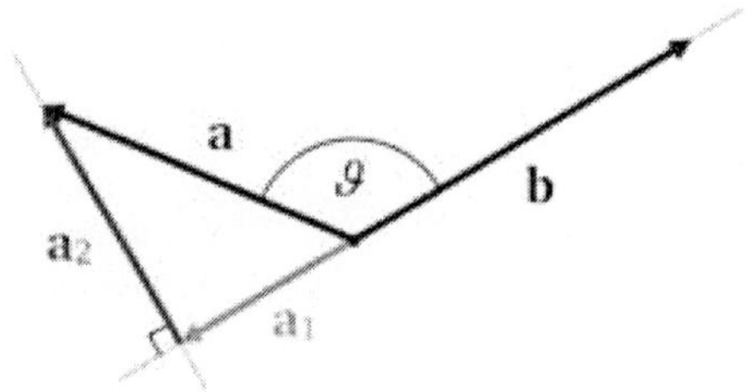

Figure: *When 90° < θ ≤ 180°, a_1 has an opposite direction with respect to b.*

Notation

Typically, a vector projection is denoted with an uppercase letter (e.g. a_1), and the corresponding scalar projection with a lowercase letter (e.g. a_1). In some cases, especially in handwriting, the vector projection is also denoted using a diacritic above or below the letter. The vector projection of a on b and the corresponding rejection are sometimes denoted by $a_{\perp b}$ and $a_{\perp b}$, respectively.

Definitions Based on Angle θ

Scalar Projection

The scalar projection of a on b is a scalar equal to

$$a_1 = |\,a\,| \cos\theta$$

where θ is the angle between a and b.

A scalar projection can be used as a scale factor to compute the corresponding vector projection.

Vector Projection

The vector projection of a on b is a vector whose magnitude is the scalar projection of a on b and whose angle against b is either 0 or 180 degrees. Namely, it is defined as

$$a_1 = a_1\hat{b} = (|a|\cos\theta)\hat{b}$$

where a_1 is the corresponding scalar projection, as defined above, and b is the unit vector with the same direction as b:

$$\hat{b} = \frac{b}{|b|}$$

Vector Rejection

By definition, the vector rejection of a on b is

$$a_2 = a - a_1$$

Hence,

$$a_2 = a - (|a|\cos\theta)\hat{b}.$$

Definitions in Terms of a and b

When θ is not known, the cosine of θ can be computed in terms of a and b, by the following property of the dot product a · b:

$$\frac{a \cdot b}{|a||b|} = \cos\theta$$

Scalar Projection

By the above mentioned property of the dot product, the definition of the scalar projection becomes

$$a_1 = |a|\cos\theta = |a|\frac{a \cdot b}{|a||b|} = \frac{a \cdot b}{|b|}$$

Vector Projection

Similarly, the definition of the vector projection becomes

$$a_1 = a_1\hat{b} = \frac{a \cdot b}{|b|}\frac{b}{|b|},$$

which is equivalent to either

$$a_1 = (a \cdot \hat{b})\hat{b},$$

or

$$a_1 = \frac{a \cdot b}{|b|^2}b = \frac{a \cdot b}{b \cdot b}b.$$

The latter formula is computationally more efficient than the former. Both require two dot products and eventually the multiplication of a scalar by a vector, but the former additionally requires a square root and the division of a vector by a scalar, while the latter additionally requires only the division of a scalar by a scalar.

Vector Rejection

By definition,

$$a_2 = a - a_1$$

Hence,

$$a_2 = a - \frac{a \cdot b}{b \cdot b} b.$$

Properties

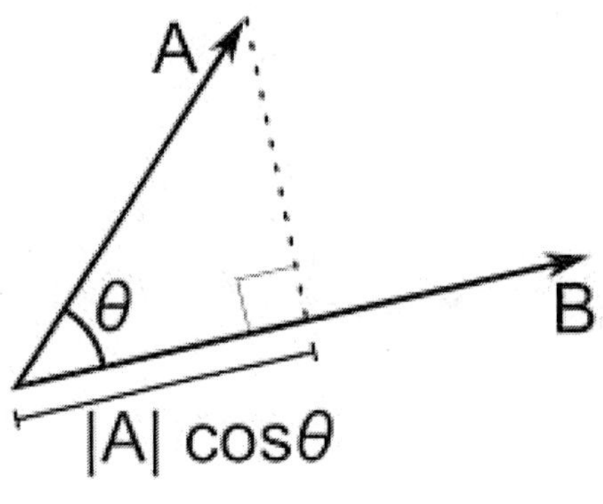

Figure: If $0° \leq \theta \leq 90°$, as in this case, the scalar projection of a on b coincides with the length of the vector projection.

Scalar Projection

The scalar projection a on b is a scalar which has a negative sign if $90 < \theta \leq 180$ degrees. It coincides with the length $|c|$ of the vector projection if the angle is smaller than 90°. More exactly:

$a_1 = |a_1|$ if $0 \leq \theta \leq 90$ degrees,

$a_1 = -|a_1|$ if $90 < \theta \leq 180$ degrees.

Vector Projection

The vector projection of a on b is a vector a_1 which is either null or parallel to b. More exactly,

- $a_1 = 0$ if $\theta = 90°$,
- a_1 and b have the same direction if $0 \leq \theta < 90$ degrees,
- a_1 and b have opposite directions if $90 < \theta \leq 180$ degrees.

Vector Rejection

The vector rejection of a on b is a vector a_2 which is either null or orthogonal to b. More exactly,

- $a_2 = 0$ if $\theta = 0$ degrees or $\theta = 180$ degrees,
- a_2 is orthogonal to b if $0 < \theta < 180$ degrees,

Matrix Representation

The orthogonal projection can be represented by a projection matrix. To project a vector onto the unit vector $a = (a_x, a_y, a_z)$, it would need to be multiplied with this projection matrix:

$$P_a = aa^T = \begin{bmatrix} a_x \\ a_y \\ a_z \end{bmatrix} \begin{bmatrix} a_x & a_y & a_z \end{bmatrix} = \begin{bmatrix} a_x^2 & a_x a_y & a_x a_z \\ a_x a_y & a_y^2 & a_y a_z \\ a_x a_z & a_y a_z & a_z^2 \end{bmatrix}$$

Uses

The vector projection is an important operation in the Gram-Schmidt orthonormalization of vector space bases. It is also used in the Separating axis theorem to detect whether two convex shapes intersect.

Generalizations

Since the notions of vector length and angle between vectors can be generalized to any n-dimensional inner product space, this is also true for the notions of orthogonal projection of a vector, projection of a vector onto another, and rejection of a vector from another. In some cases, the inner product coincides with the dot product. Whenever they don't coincide, the inner product is used instead of the dot product in the formal definitions of projection and rejection.

For a three-dimensional inner product space, the notions of projection of a vector onto another and rejection of a vector from another can be generalized to the notions of projection of a vector onto a plane, and rejection of a vector from a plane.

The projection of a vector on a plane is its orthogonal projection on that plane. The rejection of a vector from a plane is its orthogonal projection on a straight line which is orthogonal to that plane. Both are vectors. The first is parallel to the plane, the second is orthogonal. For a given vector and plane, the sum of projection and rejection is equal to the original vector.

Similarly, for inner product spaces with more than three dimensions, the notions of projection onto a vector and rejection from a vector can be generalized to the notions of projection onto a hyperplane, and rejection from a hyperplane.

In geometric algebra, they can be further generalized to the notions of projection and rejection of a general multivector onto/from any invertible k-blade.

Vector Spaces Without Fields

In mathematics, the conventional definition of the concept of vector space relies upon the algebraic concept of a field. This article treats an algebraic definition that does not require that concept. If vector spaces are redefined as (universal) algebras as below, no preliminary introduction of fields is necessary. On the contrary, fields will come from such vector space algebras.

One of the ways to do this is to keep the first four Abelian group axioms on addition in the standard formal definition and to formalize its geometric idea of scaling only by notions that concern every universal algebra. Vector space algebras consist of one binary operation + and of "unary" operations δ, which form a nonempty set Δ, that satisfy the following conditions, which do not involve fields.

1. (Total homogeneous algebra) There is a single set V such that every operation takes its two arguments or its argument from the whole V and gives its value in it.
2. (Abelian group) + satisfies the above mentioned axioms.
3. (Basis dilation) There is a basis set $B \subseteq V$ such that, for every such δ that is not constant, all the values $\delta(b)$, where b ranges over B, again form a basis set.
4. (Dilatations) Δ is the set of all and only functions that satisfy the previous conditions and preserve all

operations, namely $\delta(v+w)=\delta(v)+\delta(w)$ and $\delta(\gamma(v))=\gamma(\delta(v))$, for all $\delta,\gamma\in\Delta$ and all $v,w\in V$.

Ricci (2008) proves that these vector space algebras are the very universal algebras that any standard vector space defines by its addition and the scalar multiplications by any given scalar (namely each $a\in F$ gets a $\delta\in\Delta$ such that $\delta(v)=av$).

Ricci (2007) proves that every such a universal algebra defines a suitable field. (Hence, it proves that these conditions imply all the axioms of the standard formal definition, as well as all the defining properties in definition 3 of a field.)

Since the field is defined from such vector space algebra, this is an algebraic construction of fields, which is an instance of a more general algebraic construction: the "endowed dilatation monoid" (Ricci 2007). However, as far as fields are concerned, there also is a geometric construction. Chapter II in (Artin 1957) shows how to get them starting from geometric axioms concerning points and lines.

Graded Vector Space

In mathematics, a graded vector space is a type of vector space that includes the extra structure of gradation, which is a decomposition of the vector space into a direct sum of vector subspaces.

N-graded Vector Spaces

Let $\mathbb{N}$ be the set of non $\mathbb{N}$ -negative integers. An -graded vector space $\mathbb{N}$, often called simply a graded vector space without the prefix , is a vector space V which decomposes into a direct sum of the form

$$V=\bigoplus_{n\in\mathbb{N}}V_n$$

where each V_n is a vector space. For a given n the elements of V_n are then called homogeneous elements of degree n.

Graded vector spaces are common. For example the set of all polynomials in one variable form a graded vector space, where the homogeneous elements of degree n are exactly the linear combinations of monomials of degree n.

General I-graded Vector Spaces

The subspaces of a graded vector space need not be indexed by the set of natural numbers, and may be indexed by the elements of any set I. An I-graded vector space V is a vector space that can be written as a direct sum of subspaces indexed by elements i of set I:

$$V = \bigoplus_{i \in I} V_i .$$

Therefore, an $\mathbb{N}$-graded vector space, as defined above, is just an I-graded vector space where the set I is $\mathbb{N}$ (the set of natural numbers).

The case where I is the ring $\mathbb{Z}/2\mathbb{Z}$ (the elements 0 and 1) is particularly important in physics. A $(\mathbb{Z}/2\mathbb{Z})$-graded vector space is also known as a supervector space.

Linear Maps

For general index sets I, a linear map between two I-graded vector spaces $f: V \rightarrow W$ is called a graded linear map if it preserves the grading of homogeneous elements:

$$f(V_i) \subseteq W_i \text{ for all i in } I.$$

When I is a commutative monoid (such as the natural numbers), then one may more generally define linear maps that are homogeneous of any degree i in I by the property

$$f(V_j) \subseteq W_{i+j} \text{ for all j in } I,$$

where "+" denotes the monoid operation. If moreover I satisfies the cancellation property so that it can be embedded into a commutative group A which it generates (for instance the integers if I is the natural numbers), then one may also define linear maps that are homogeneous of degree i in A by the same property (but now "+" denotes the group operation in A). In particular for i in I a linear map will be homogeneous of degree –i if

$$f(V_{i+j}) \subseteq W_j \text{ for all j in I, while}$$

$$f(V_j) = 0 \text{ if j–i is not in I.}$$

Just as the set of linear maps from a vector space to itself forms an associative algebra (the algebra of endomorphisms of the vector space), the sets of homogeneous linear maps from a space to itself, either restricting degrees to I or allowing any degrees in the group A, form associative graded algebras over those index sets.

Operations on Graded Vector Spaces

Some operations on vector spaces can be defined for graded vector spaces as well. Given two I-graded vector spaces V and W, their direct sum has underlying vector space $V \oplus W$ with gradation

$$(V \oplus W)_i = V_i \oplus W_i .$$

If I is a semigroup, then the tensor product of two I-graded vector spaces V and W is another I-graded vector space, $V \otimes W$ with gradation

$$(V \otimes W)_i = \bigoplus_{\{j,k|j+k=i\}} V_j \otimes W_k .$$

Vector-valued Function

A vector-valued function, also referred to as a vector function, is a mathematical function of one or more variables whose range is a set of multidimensional vectors or infinite-dimensional vectors. The input of a vector-valued function could be a scalar or a vector. The dimension of the domain is not defined by the dimension of the range.

Properties

The domain of a vector-valued function is the intersection of the domain of the functions f, g, and h.

Derivative of a Three-dimensional Vector Function

Many vector-valued functions, like scalar-valued functions, can be differentiated by simply differentiating the components in the Cartesian coordinate system. Thus, if

$$\mathrm{r}(t) = f(t)\mathrm{i} + g(t)\mathrm{j} + h(t)\mathrm{k}$$

is a vector-valued function, then

$$\frac{d\mathrm{r}(t)}{dt} = f'(t)\mathrm{i} + g'(t)\mathrm{j} + h'(t)\mathrm{k}.$$

The vector derivative admits the following physical interpretation: if r(t) represents the position of a particle, then the derivative is the velocity of the particle

$$\mathrm{v}(t) = \frac{d\mathrm{r}(t)}{dt}.$$

Likewise, the derivative of the velocity is the acceleration

$$\frac{d\mathrm{v}(t)}{dt} = \mathrm{a}(t).$$

Partial Derivative

The partial derivative of a vector function a with respect to a scalar variable q is defined as

$$\frac{\partial \mathrm{a}}{\partial q} = \sum_{i=1}^{3} \frac{\partial a_i}{\partial q} \mathrm{e}_i$$

where a_i is the scalar component of a in the direction of e_i. It is also called the direction cosine of a and e_i or their dot product. The vectors e_1, e_2, e_3 form an orthonormal basis fixed in the reference frame in which the derivative is being taken.

Ordinary Derivative

If a is regarded as a vector function of a single scalar variable, such as time t, then the equation above reduces to the first ordinary time derivative of a with respect to t,

$$\frac{d\mathrm{a}}{dt} = \sum_{i=1}^{3} \frac{da_i}{dt} \mathrm{e}_i.$$

Total Derivative

If the vector a is a function of a number n of scalar variables q_r (r = 1,...,n), and each q_r is only a function of time t, then the ordinary derivative of a with respect to t can be expressed, in a form known as the total derivative, as

$$\frac{d\mathrm{a}}{dt} = \sum_{r=1}^{n} \frac{\partial \mathrm{a}}{\partial q_r} \frac{dq_r}{dt} + \frac{\partial \mathrm{a}}{\partial t}.$$

Some authors prefer to use capital *D* to indicate the total derivative operator, as in *D*/*D*t. The total derivative differs from the partial time derivative in that the total derivative accounts for changes in a due to the time variance of the variables q_r.

Reference Frames

Whereas for scalar-valued functions there is only a single possible reference frame, to take the derivative of a vector-valued function requires the choice of a reference frame (at least when a fixed Cartesian coordinate system is not implied as such). Once a reference frame has been chosen, the derivative of a vector-valued function can be computed using techniques similar to those for computing derivatives of scalar-valued functions. A different choice of reference frame will, in general, produce a different derivative function. The derivative functions in different reference frames have a specific kinematical relationship.

Derivative of a Vector Function with Nonfixed Bases

The above formulas for the derivative of a vector function rely on the assumption that the basis vectors e_1,e_2,e_3 are constant, that is, fixed in the reference frame in which the derivative of a is being taken, and therefore the e_1,e_2,e_3 each has a derivative of identically zero. This often holds true for problems dealing with vector fields in a fixed coordinate system, or for simple problems in physics. However, many complex problems involve the derivative of a vector function in multiple moving reference frames, which means that the basis vectors will not necessarily be constant. In such a case where the basis vectors e_1,e_2,e_3 are fixed in reference frame E, but not in reference frame N, the more general formula for the ordinary time derivative of a vector in reference frame N is

$$\frac{{}^{\mathrm{N}}da}{dt} = \sum_{i=1}^{3} \frac{da_i}{dt} \mathrm{e}_i + \sum_{i=1}^{3} a_i \frac{{}^{\mathrm{N}}de_i}{dt}$$

where the superscript N to the left of the derivative operator indicates the reference frame in which the derivative is taken. As shown previously, the first term on the right hand side is equal to the derivative of a in the reference frame where e_1,e_2,e_3 are constant, reference frame E. It also can be shown that the second term on the right hand side is equal to the relative angular velocity of the two reference frames cross multiplied with the vector a itself. Thus, after substitution, the formula relating the derivative of a vector function in two reference frames is

$$\frac{{}^{\mathrm{N}}da}{dt} = \frac{{}^{\mathrm{E}}da}{dt} + {}^{\mathrm{N}}\omega^{\mathrm{E}} \times \mathrm{a}$$

where ${}^{N}\omega^{E}$ is the angular velocity of the reference frame E relative to the reference frame N.

One common example where this formula is used is to find the velocity of a space-borne object, such as a rocket, in the inertial reference frame using measurements of the rocket's velocity relative to the ground. The velocity ${}^{N}v^{R}$ in inertial reference frame N of a rocket R located at position r^{R} can be found using the formula

$$\frac{{}^{N}d}{dt}(r^{R}) = \frac{{}^{E}d}{dt}(r^{R}) + {}^{N}\omega^{E} \times r^{R}.$$

where ${}^{N}\omega^{E}$ is the angular velocity of the Earth relative to the inertial frame N. Since velocity is the derivative of position, ${}^{N}v^{R}$ and ${}^{E}v^{R}$ are the derivatives of r^{R} in reference frames N and E, respectively. By substitution,

$${}^{N}v^{R} = {}^{E}v^{R} + {}^{N}\omega^{E} \times r^{R}$$

where ${}^{E}v^{R}$ is the velocity vector of the rocket as measured from a reference frame E that is fixed to the Earth.

Derivative and Vector Multiplication

The derivative of the products of vector functions behaves similarly to the derivative of the products of scalar functions. Specifically, in the case of scalar multiplication of a vector, if p is a scalar variable function of q,

$$\frac{\partial}{\partial q}(p\mathrm{a}) = \frac{\partial p}{\partial q}\mathrm{a} + p\frac{\partial \mathrm{a}}{\partial q}.$$

In the case of dot multiplication, for two vectors a and b that are both functions of q,

$$\frac{\partial}{\partial q}(\mathrm{a}\cdot\mathrm{b}) = \frac{\partial \mathrm{a}}{\partial q}\cdot\mathrm{b} + \mathrm{a}\cdot\frac{\partial \mathrm{b}}{\partial q}.$$

Similarly, the derivative of the cross product of two vector functions is

$$\frac{\partial}{\partial q}(\mathrm{a}\times\mathrm{b}) = \frac{\partial \mathrm{a}}{\partial q}\times\mathrm{b} + \mathrm{a}\times\frac{\partial \mathrm{b}}{\partial q}.$$

Derivative of an n-dimensional Vector Function

A function f of a real number t with values in the space R^n can be written as $f(t) = (f_1(t), f_2(t), \ldots, f_n(t))$. Its derivative equals

$$f'(t) = (f_{1'}(t), f_{2'}(t), \ldots, f_{n'}(t)).$$

If f is a function of several variables, say of $t \in R^m$, then the partial derivatives of the components of f form a $n \times m$ matrix called the Jacobian matrix of f.

Infinite-dimensional Vector Functions

If the values of a function f lie in an infinite-dimensional vector space *X*, such as a Hilbert space, then *f* may be called an infinite-dimensional vector function.

Functions with Values in a Hilbert Space

If the argument of f is a real number and *X* is a Hilbert space, then the derivative of f at a point t can be defined as in the finite-dimensional case:

$$f'(t) = \lim_{h \to 0} \frac{f(t+h) - f(t)}{h}.$$

Most results of the finite-dimensional case also hold in the infinite-dimensional case too, mutatis mutandis. Differentiation can also be defined to functions of several variables (e.g. $t \in R^n$, or even $t \in Y$, where *Y* is an infinite-dimensional vector space).

N.B. If X is a Hilbert space, then one can easily show that any derivative (and any other limit) can be computed componentwise: if

$$f = (f_1, f_2, f_3, \ldots)$$

(i.e., $f = f_1 e_1 + f_2 e_2 + f_3 e_3 + \cdots$, where $e_1, e_2, e_3, \ldots$ is an orthonormal basis of the space X), and $f'(t)$ exists, then

$$f'(t) = (f_{1'}(t), f_{2'}(t), f_{3'}(t), \ldots).$$

However, the existence of a componentwise derivative does not guarantee the existence of a derivative, as componentwise convergence in a Hilbert space does not guarantee convergence with respect to the actual topology of the Hilbert space.

Other Infinite-dimensional Vector Spaces

Most of the above hold for other topological vector spaces *X* too. However, not as many classical results hold in the Banach space setting, e.g., an absolutely continuous function with values in a suitable Banach space need not have a derivative anywhere. Moreover, in most Banach spaces setting there are no orthonormal bases.

Wave Vector

In physics, a wave vector (also spelled wavevector) is a vector which helps describe a wave. Like any vector, it has a magnitude and direction, both of which are important: Its magnitude is either the wavenumber or angular wavenumber of the wave (inversely proportional to the wavelength), and its direction is ordinarily the direction of wave propagation.

In the context of special relativity the wave vector can also be defined as a four-vector.

Definitions

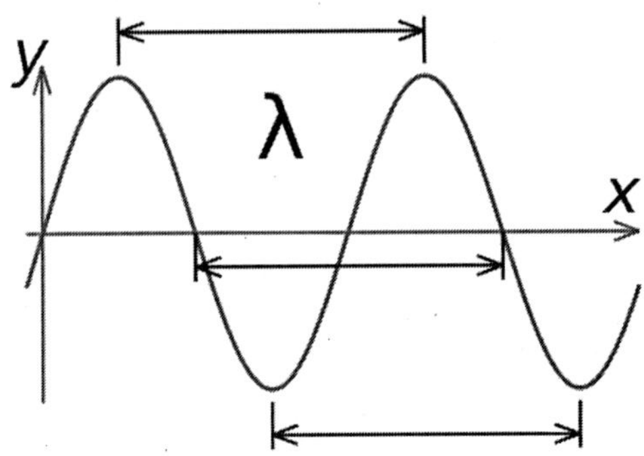

Figure: *Wavelength of a sine wave, λ, can be measured between any two consecutive points with the same phase, such as between adjacent crests, or troughs, or adjacent zero crossings with the same direction of transit, as shown.*

Unfortunately, there are two common definitions of wave vector, which differ by a factor of 2π in their magnitudes. One definition is preferred in physics and related fields, while the other definition is preferred in crystallography and related fields. For this article, they will be called the "physics definition" and the "crystallography definition", respectively.

Physics Definition

A perfect one-dimensional travelling wave follows the equation:

$$\psi(x,t) = A\cos(kx - \omega t + \varphi)$$

where:

- x is position,
- t is time,
- ψ (a function of x and t) is the disturbance describing the wave (for example, for an ocean wave, ψ would be the excess height of the water, or for a sound wave, ψ would be the excess air pressure).
- A is the amplitude of the wave (the peak magnitude of the oscillation),
- φ is a "phase offset" describing how two waves can be out of sync with each other,
- ω is the angular frequency of the wave, related to how quickly it oscillates at a given point,
- k is the wavenumber (more specifically called angular wavenumber) of the wave, related to the wavelength by the equation $k = 2\pi / \lambda$.

This wave travels in the $+x$ direction with speed (more specifically, phase velocity) ω / k.

Crystallography Definition

In crystallography, the same waves are described using slightly different equations. In one and three dimensions respectively:

$$\psi(x,t) = A\cos(2\pi(kx - \nu t) + \varphi)$$

$$\psi(\mathrm{r},t) = A\cos\left(2\pi(\mathrm{k}\cdot\mathrm{r} - \nu t) + \varphi\right)$$

The differences are:

- The frequency ν instead of angular frequency ω is used. They are related by $2\pi\nu = \omega$. This substitution is not important for this article, but reflects common practice in crystallography.
- The wavenumber k and wave vector k are defined in a different way. Here, $k = |\mathrm{k}| = 1/\lambda$, while in the physics definition above, $k = |\mathrm{k}| = 2\pi/\lambda$. The direction of k is discussed below.

Direction of the Wave Vector

The direction in which the wave vector points must be distinguished from the "direction of wave propagation". The "direction of wave propagation" is the direction of a wave's energy flow, and the direction that a small wave packet will move, i.e. the direction of the group velocity. For light waves, this is also the direction of the Poynting vector. On the other hand, the wave vector points in the direction of phase velocity. In other words, the wave vector points in the normal direction to the surfaces of constant phase, also called wave fronts.

In a lossless isotropic medium such as air, any gas, any liquid, or some solids (such as glass), the direction of the wavevector is exactly the same as the direction of wave propagation. If the medium is lossy, the wave vector in general points in directions other than that of wave propagation. The condition for wave vector to point in the same direction in which the wave propagates is that the wave has to be homogeneous, which isn't necessarily satisfied when the medium is lossy. In a homogeneous wave, the surfaces of constant phase are also surfaces of constant amplitude. In case of inhomogeneous waves, these two species of surfaces differ in orientation. Wave vector is always perpendicular to surfaces of constant phase.

However, when a wave travels through an anisotropic medium, such as light waves through an asymmetric crystal or sound waves through a sedimentary rock, the wave vector may not point exactly in the direction of wave propagation.

In Solid-state Physics

In solid-state physics, the "wavevector" (also called k-vector) of an electron or hole in a crystal is the wavevector of its quantum-mechanical wavefunction. These electron waves are not ordinary sinusoidal waves, but they do have a kind of envelope function which is sinusoidal, and the wavevector is defined via that envelope wave, usually using the "physics definition".

In Special Relativity

A beam of coherent, monochromatic light can be characterized by the (null) wave 4-vector

$$k^{\mu} = \left(\frac{\omega}{c}, \vec{k}\right)$$

which, when written out explicitly in its contravariant and covariant forms is

$$k^{\mu} = \left(\frac{\omega}{c}, k_x, k_y, k_z\right) \text{ and}$$

$$k_{\mu} = \left(\frac{\omega}{c}, -k_x, -k_y, -k_z\right).$$

The null character of the wave 4-vector gives a relation between the frequency and the magnitude of the spatial part of the wave 4-vector:

$$k^{\mu}k_{\mu} = \left(\frac{\omega}{c}\right)^2 - k_x^2 - k_y^2 - k_z^2 = 0$$

Lorentz Transformation

Taking the Lorentz transformation of the wave vector is one way to derive the relativistic Doppler effect. The Lorentz matrix is defined as

$$\Lambda = \begin{pmatrix} \gamma & -\beta\gamma & 0 & 0 \\ -\beta\gamma & \gamma & 0 & 0 \\ 0 & 0 & 1 & 0 \\ 0 & 0 & 0 & 1 \end{pmatrix}$$

In the situation where light is being emitted by a fast moving source and one would like to know the frequency of light detected in an earth (lab) frame, we would apply the Lorentz transformation as follows. Note that the source is in a frame S^s and earth is in the observing frame, S^{obs}. Applying the lorentz transformation to the wave vector

$$k_s^{\mu} = \Lambda^{\mu}_{\nu} k^{\nu}_{\text{obs}}$$

and choosing just to look at the $\mu = 0$ component results in

$$k_s^0 = \Lambda^0_0 k^0_{\text{obs}} + \Lambda^0_1 k^1_{\text{obs}} + \Lambda^0_2 k^2_{\text{obs}} + \Lambda^0_3 k^3_{\text{obs}}$$

$$\frac{\omega_s}{c} = \gamma\frac{\omega_{\text{obs}}}{c} - \beta\gamma k^1_{\text{obs}}$$

$$= \gamma \frac{\omega_{\text{obs}}}{c} - \beta\gamma \frac{\omega_{\text{obs}}}{c} \cos\theta.$$

where $\cos\theta$ is the direction cosine of k^1 wrt $k^0, k^1 = k^0 \cos\theta$.

So

$$\frac{\omega_{\text{obs}}}{\omega_s} = \frac{1}{\gamma(1 - \beta\cos\theta)}$$

Source Moving Away

As an example, to apply this to a situation where the source is moving directly away from the observer ($\theta = \pi$), this becomes:

$$\frac{\omega_{\text{obs}}}{\omega_s} = \frac{1}{\gamma(1+\beta)} = \frac{\sqrt{1-\beta^2}}{1+\beta} = \frac{\sqrt{(1+\beta)(1-\beta)}}{1+\beta} = \frac{\sqrt{1-\beta}}{\sqrt{1+\beta}}$$

Source Moving Towards

To apply this to a situation where the source is moving straight towards the observer ($\theta = \pi$), this becomes

$$\frac{\omega_{\text{obs}}}{\omega_s} = \frac{\sqrt{1+\beta}}{\sqrt{1-\beta}}$$

Vectorization (Mathematics)

In mathematics, especially in linear algebra and matrix theory, the vectorization of a matrix is a linear transformation which converts the matrix into a column vector.

Specifically, the vectorization of an m×n matrix A, denoted by vec(A), is the mn × 1 column vector obtained by stacking the columns of the matrix A on top of one another:

$$\text{vec}(A) = [a_{1,1}, \ldots, a_{m,1}, a_{1,2}, \ldots, a_{m,2}, \ldots, a_{1,n}, \ldots, a_{m,n}]^T$$

Here $a_{i,j}$ represents the (i, j)-th element of matrix A and the superscript$T$denotes the transpose.

Vectorization expresses the isomorphism $\mathrm{R}^{m \times n} := \mathrm{R}^m \otimes \mathrm{R}^n \cong \mathrm{R}^{mn}$ between these vector spaces (of matrices and vectors) in coordinates.

For example, for the 2×2 matrix = $\begin{bmatrix} a & b \\ c & d \end{bmatrix}$, the vectorization is $\text{vec}(A) = \begin{bmatrix} a \\ c \\ b \\ d \end{bmatrix}$.

Compatibility with Kronecker Products

The vectorization is frequently used together with the Kronecker product to express matrix multiplication as a linear transformation on matrices. In particular,

$$\text{vec}(ABC) = (C^T \otimes A)\text{vec}(B)$$

for matrices A, B, and C of dimensions k×l, l×m, and m×n. For example, if (the adjoint endomorphism of the Lie algebra gl(n,C) of all n×n matrices with complex entries), then $\text{vec}(\text{ad}_A(X)) = (I_n \otimes A - A^T \otimes I_n)\text{vec}(X)$, where I_n is the n×n identity matrix. There are two other useful formulations:

$$\text{vec}(ABC) = (I_n \otimes AB)\text{vec}(C) = (C^T B^T \otimes I_k)\text{vec}(A)$$

$$\text{vec}(AB) = (I_m \otimes A)\text{vec}(B) = (B^T \otimes I_k)\text{vec}(A)$$

Compatibility with Hadamard Products

Vectorization is an algebra homomorphism from the space of n×n matrices with the Hadamard (entrywise) product to C^n with its Hadamard product:

$$\text{vec}(A \circ B) = \text{vec}(A) \circ \text{vec}(B).$$

Compatibility with Inner Products

Vectorization is a unitary transformation from the space of n×n matrices with the Frobenius (or Hilbert-Schmidt) inner product to C^n :

$$\text{tr}(A^* B) = \text{vec}(A)^* \text{vec}(B)$$

where the superscript * denotes the conjugate transpose.

Half-vectorization

For a symmetric matrix A, the vector vec(A) contains more information than is strictly necessary, since the matrix is completely determined by the symmetry together with the lower triangular portion, that is, the n(n+1)/2 entries on and below the main diagonal. For such matrices, the half-vectorization is sometimes more useful than the vectorization. The half-vectorization, vech(A), of a symmetric n×n matrix A is the n(n+1)/2 × 1 column vector obtained by vectorizing only the lower triangular part of A:

$$\text{vech}(A) = [\,A_{1,1}, \ldots, A_{m,1}, A_{2,2}, \ldots, A_{n,2}, \ldots, A_{n-1,n-1}, A_{n-1,n}, A_{n,n}\,]^T.$$

For example, for the 2×2 matrix $A = \begin{bmatrix} a & b \\ b & d \end{bmatrix}$, the half-vectorization is $\text{vech}(A) = \begin{bmatrix} a \\ b \\ d \end{bmatrix}$.

There exist unique matrices transforming the half-vectorization of a matrix to its vectorization and vice-versa called, respectively, the duplication matrix and the elimination matrix.

Programming Language

Programming languages that implement matrices may have easy means for vectorization. In Matlab/GNU Octave a matrix *A* can be vectorized by *A*(:). In Python NumPy arrays implement the 'flatten' method (although this stacks the rows of the matrix, not the columns), while in *R* the desired effect can be achieved via the '*c*()' or 'as.vector()' functions.

Chapter 5

Euclidean Vector

This article is about the vectors mainly used in physics and engineering to represent directed quantities.

In mathematics, physics, and engineering, a Euclidean vector (sometimes called a geometric or spatial vector, or – as here – simply a vector) is a geometric object that has a magnitude (or length) and direction and can be added according to the parallelogram law of addition. A Euclidean vector is frequently represented by a line segment with a definite direction, or graphically as an arrow, connecting an *initial point A* with a *terminal point B*, and denoted by $\overrightarrow{AB}$.

Vectors play an important role in physics: velocity and acceleration of a moving object and forces acting on it are all described by vectors. Many other physical quantities can be usefully thought of as vectors. Although most of them do not represent distances (except, for example, position or displacement), their magnitude and direction can be still represented by the length and direction of an arrow. The mathematical representation of a physical vector depends on the coordinate system used to describe it. Other vector-like objects that describe physical quantities and transform in a similar way under changes of the coordinate system include pseudovectors and tensors.

It is important to distinguish Euclidean vectors from the more general concept in linear algebra of vectors as elements of a vector space. General vectors in this sense are fixed-size, ordered collections of items as in the case of Euclidean vectors, but the individual items may not be real numbers, and the normal Euclidean concepts of length, distance and angle may not be applicable. (A vector space with a definition of these concepts is called an inner product space.) In turn, both of these definitions of vector should be distinguished from the statistical concept of a random vector. The individual items in a random vector are individual real-valued random variables, and are often manipulated using the same sort of mathematical vector and matrix operations that apply to the other types of vectors, but otherwise usually behave more like collections of individual values. Concepts of length, distance and angle do not normally apply to these vectors, either; rather, what links the values together is the potential correlations among them.

Overview

A vector is a geometric entity characterized by a magnitude (in mathematics a number, in physics a number times a unit) and a direction. In rigorous mathematical treatments, a vector is defined as a directed line segment, or arrow, in a Euclidean space. When it becomes necessary to distinguish it from vectors as defined elsewhere, this is sometimes referred to as a geometric, spatial, or Euclidean vector.

As an arrow in Euclidean space, a vector possesses a definite *initial point* and *terminal point*. Such a vector is called a bound vector. When only the magnitude and direction of the vector matter, then the particular initial point is of no importance, and the vector is called a free vector. Thus two arrows $\overrightarrow{AB}$ and $\overrightarrow{A'B'}$ in space represent the same free vector if they have the same magnitude and direction: that is, they are equivalent if the quadrilateral $ABB2A2$ is a parallelogram. If the Euclidean space is equipped with a choice of origin, then a free vector is equivalent to the bound vector of the same magnitude and direction whose initial point is the origin.

The term *vector* also has generalizations to higher dimensions and to more formal approaches with much wider applications.

Examples in One Dimension

Since the physicist's concept of force has a direction and a magnitude, it may be seen as a vector. As an example, consider a rightward force F of 15 newtons. If the positive axis is also directed rightward, then F is represented by the vector 15 N, and if positive points leftward, then the vector for F is -15 N. In either case, the magnitude of the vector is 15 N.

Likewise, the vector representation of a displacement Δs of 4 metres to the right would be 4 m or -4 m, and its magnitude would be 4 m regardless.

In Physics and Engineering

Vectors are fundamental in the physical sciences. They can be used to represent any quantity that has both a magnitude and direction, such as velocity, the magnitude of which is speed. For example, the velocity *5 metres per second upward* could be represented by the vector (0,5) (in 2 dimensions with the positive y axis as 'up').

Another quantity represented by a vector is force, since it has a magnitude and direction. Vectors also describe many other physical quantities, such as displacement, acceleration, momentum, and angular momentum.

Other physical vectors, such as the electric and magnetic field, are represented as a system of vectors at each point of a physical space; that is, a vector field.

In Cartesian Space

In the Cartesian coordinate system, a vector can be represented by identifying the coordinates of its initial and terminal point. For instance, the points A = (1,0,0) and B = (0,1,0) in space determine the free vector $\overrightarrow{AB}$ pointing from the point x=1 on the x-axis to the point y=1 on the y-axis.

Typically in Cartesian coordinates, one considers primarily bound vectors. A bound vector is determined by the coordinates of the terminal point, its initial point always having the coordinates of the origin O = (0,0,0). Thus the bound vector represented by (1,0,0) is a vector of unit length pointing from the origin up the positive x-axis.

The coordinate representation of vectors allows the algebraic features of vectors to be expressed in a convenient numerical fashion. For example, the sum of the vectors (1,2,3) and (-2,0,4) is the vector

$$(1, 2, 3) + (-2, 0, 4) = (1 - 2, 2 + 0, 3 + 4) = (-1, 2, 7).$$

Euclidean and Affine Vectors

In the geometrical and physical settings, sometimes it is possible to associate, in a natural way, a *length* or magnitude and a direction to vectors. In turn, the notion of direction is strictly associated with the notion of an *angle* between two vectors. When the length of vectors is defined, it is possible to also define a dot product — a scalar-valued product of two vectors — which gives a convenient algebraic characterization of both length (the square root of the dot product of a vector by itself) and angle (a function of the dot product between any two vectors). In three dimensions, it is further possible to define a cross product which supplies an algebraic characterization of the area and orientation in space of the parallelogram defined by two vectors (used as sides of the parallelogram).

However, it is not always possible or desirable to define the length of a vector in a natural way. This more general type of spatial vector is the subject of vector spaces (for bound vectors) and affine spaces (for free vectors). An important example is Minkowski space that is important to our understanding of special relativity, where there is a generalization of length that permits non-zero vectors to have zero length. Other physical examples come from thermodynamics, where many of the quantities of interest can be considered vectors in a space with no notion of length or angle.

Generalizations

In physics, as well as mathematics, a vector is often identified with a tuple, or list of numbers, which depend on some auxiliary coordinate system or reference frame. When the coordinates are transformed, for example by rotation or stretching, then the components of the vector also transform. The vector itself has not changed, but the reference frame has, so the components of the vector (or measurements taken with respect to the reference frame) must change to compensate.

The vector is called *covariant* or *contravariant* depending on how the transformation of the vector's components is related to the transformation of coordinates. In general, contravariant vectors are "regular vectors" with units of distance (such as a displacement) or distance times some other unit (such as velocity or acceleration); covariant vectors, on the other hand, have units of one-over-distance such as gradient.

If you change units (a special case of a change of coordinates) from metres to millimetres, a scale factor of 1/1000, a displacement of 1 m becomes 1000 mm–a contravariant change in numerical value. In contrast, a gradient of 1 K/m becomes 0.001 K/mm–a covariant change in value. Tensors are another type of quantity that behave in this way; in fact a vector is a special type of tensor.

In pure mathematics, a vector is any element of a vector space over some field and is often represented as a coordinate vector. The vectors described in this article are a very special case of this general definition because they are contravariant with respect to the ambient space. Contravariance captures the physical intuition behind the idea that a vector has "magnitude and direction".

History

The concept of vector, as we know it today, evolved gradually over a period of more than 200 years. About a dozen people made significant contributions. The immediate predecessor of vectors were quaternions, devised by William Rowan Hamilton in 1843 as a generalization of complex numbers.

His search was for a formalism to enable the analysis of three-dimensional space in the same way that complex numbers had enabled analysis of two-dimensional space. In 1846 Hamilton divided his quaternions into the sum of real and imaginary parts that he respectively called "scalar" and "vector":

> *The algebraically imaginary part, being geometrically constructed by a straight line, or radius vector, which has, in general, for each determined quaternion, a determined length and determined direction in space, may be called the vector part, or simply the vector of the quaternion.*
>
> *—W. R. Hamilton, London, Edinburgh & Dublin Philosophical Magazine 3rd series 29 27 (1846)*

Whereas complex numbers have one number (i) whose square is negative one, quaternions have three independent such numbers (i, j, k). Multiplication of these numbers by each other is not commutative, e.g., $ij = -ji = k$. Multiplication of two quaternions yields a third quaternion whose scalar part is the negative of the modern dot product and whose vector part is the modern cross product.

Peter Guthrie Tait carried the quaternion standard after Hamilton. His 1867 *Elementary Treatise of Quaternions* included extensive treatment of the nabla or del operator and is very close to modern vector analysis.

Josiah Willard Gibbs, who was exposed to quaternions through James Clerk Maxwell's *Treatise on Electricity and Magnetism*, separated off their vector part for independent treatment. The first half of Gibbs's *Elements of Vector Analysis*, published in 1881, presents what is essentially the modern system of vector analysis.

Representations

Vectors are usually denoted in lowercase boldface, as a or lowercase italic boldface, $\vec{a}$ as *a*. (Uppercase letters are typically used to represent matrices.) Other conventions include or *a*, especially in handwriting.

Alternatively, some use a tilde (~) or a wavy underline drawn beneath the symbol, which is a convention for indicating boldface type. If the vector represents a directed distance or displacement from a point *A* to a point *B*, it can also be denoted $\overrightarrow{AB}$ as or *AB*.

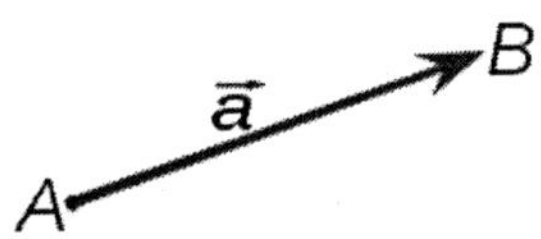

Vectors are usually shown in graphs or other diagrams as arrows (directed line segments). Here the point *A* is called the *origin*, *tail*, *base*, or *initial point*; point *B* is called the *head*, *tip*, *endpoint*, *terminal point* or *final point*. The length of the arrow is proportional to the vector's magnitude, while the direction in which the arrow points indicates the vector's direction.

On a two-dimensional diagram, sometimes a vector perpendicular to the plane of the diagram is desired. These vectors are commonly shown as small circles.

A circle with a dot at its centre (Unicode U+2299 ⊙) indicates a vector pointing out of the front of the diagram, toward the viewer.

A circle with a cross inscribed in it (Unicode U+2297 ⊗) indicates a vector pointing into and behind the diagram. These can be thought of as viewing the tip of an arrow head on and viewing the vanes of an arrow from the back.

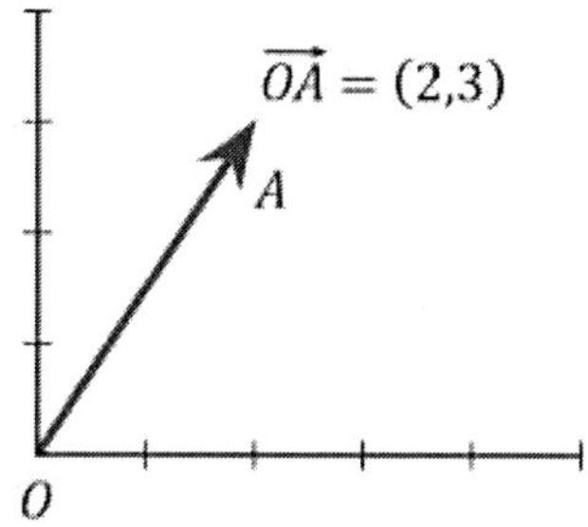

***Figure:** A vector in the Cartesian plane, showing the position of a point A with coordinates (2,3).*

In order to calculate with vectors, the graphical representation may be too cumbersome.

Vectors in an n-dimensional Euclidean space can be represented as coordinate vectors in a Cartesian coordinate system.

The endpoint of a vector can be identified with an ordered list of n real numbers (n-tuple).

These numbers are the coordinates of the endpoint of the vector, with respect to a given Cartesian coordinate system, and are typically called the scalar components (or scalar projections) of the vector on the axes of the coordinate system.

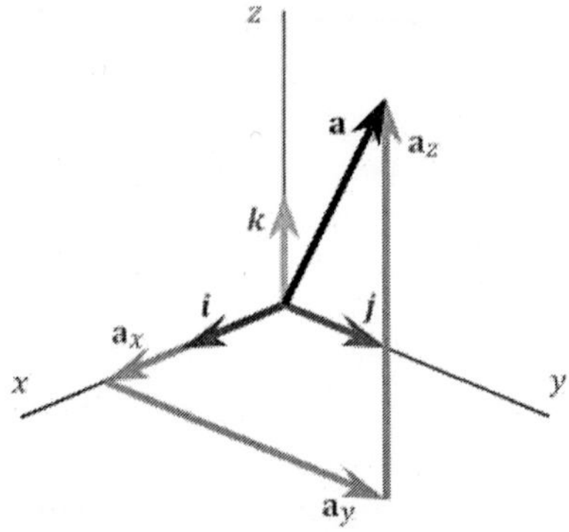

As an example in two dimensions, the vector from the origin $O = (0,0)$ to the point $A = (2,3)$ is simply written as

$$\mathbf{a} = (2,3).$$

The notion that the tail of the vector coincides with the origin is implicit and easily understood. Thus, the more explicit notation $\overrightarrow{OA}$ is usually not deemed necessary and very rarely used.

In three dimensional Euclidean space (or $\mathbb{R}^3$), vectors are identified with triples of scalar components:

$$\mathbf{a} = (a_1, a_2, a_3).$$

also written

$$\mathbf{a} = (a_x, a_y, a_z).$$

These numbers are often arranged into a column vector or row vector, particularly when dealing with matrices, as follows:

$$\mathbf{a} = \begin{bmatrix} a_1 \\ a_2 \\ a_3 \end{bmatrix}$$

$$\mathbf{a} = [a_1 \ a_2 \ a_3].$$

Another way to represent a vector in n-dimensions is to introduce the standard basis vectors. For instance, in three dimensions, there are three of them:

$$\mathbf{e}_1 = (1,0,0), \mathbf{e}_2 = (0,1,0), \mathbf{e}_3 = (0,0,1).$$

These have the intuitive interpretation as vectors of unit length pointing up the *x*, *y*, and *z* axis of a Cartesian coordinate system, respectively, and they are sometimes referred to as versors of those axes. In terms of these, any vector a in $\mathbb{R}^3$ can be expressed in the form:

or $$\mathbf{a} = \mathbf{a}_1 + \mathbf{a}_2 + \mathbf{a}_3 = a_1\mathbf{e}_1 + a_2\mathbf{e}_2 + a_3\mathbf{e}_3,$$

where a_1, a_2, a_3 are called the vector components (or vector projections) of a on the basis vectors or, equivalently, on the corresponding Cartesian axes *x*, *y*, and *z*, while a_1, a_2, a_3 are the respective scalar components (or scalar projections).

In introductory physics textbooks, the standard basis vectors are often instead denoted $\mathbf{i}, \mathbf{j}, \mathbf{k}$ (or , in which the hat symbol ^ typically denotes unit vectors). $\hat{\mathbf{x}}, \hat{\mathbf{y}}, \hat{\mathbf{z}}$ In this case, the scalar and vector components are denoted a_x, a_y, a_z, and a_x, a_y, a_z. Thus,

$$\mathbf{a} = \mathbf{a}_x + \mathbf{a}_y + \mathbf{a}_z = a_x\mathbf{i} + a_y\mathbf{j} + a_z\mathbf{k}.$$

The notation e_i is compatible with the index notation and the summation convention commonly used in higher level mathematics, physics, and engineering.

Decomposition

As explained above a vector is often described by a set of vector components that are mutually perpendicular and add up to form the given vector. Typically, these components are the projections of the vector on a set of reference axes (or basis vectors). The vector is said to be *decomposed* or *resolved with respect to* that set.

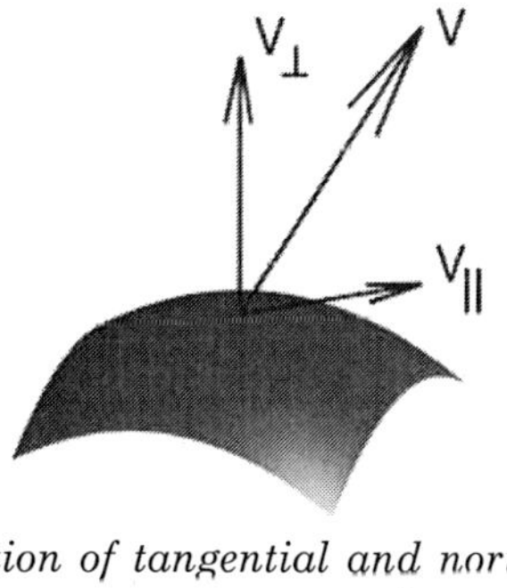

Figure: *Illustration of tangential and normal components of a vector to a surface.*

However, the decomposition of a vector into components is not unique, because it depends on the choice of the axes on which the vector is projected.

Moreover, the use of Cartesian versors such as $\hat{\mathbf{x}}, \hat{\mathbf{y}}, \hat{\mathbf{z}}$ as a basis in which to represent a vector is not mandated. Vectors can also be expressed in terms of the versors of a Cylindrical coordinate system $(\hat{\boldsymbol{\rho}}, \hat{\phi}, \hat{\mathbf{z}})$ or Spherical coordinate system $(\hat{\mathbf{r}}, \hat{\boldsymbol{\theta}}, \hat{\phi})$. The latter two choices are more convenient for solving problems which possess cylindrical or spherical symmetry respectively.

The choice of a coordinate system doesn't affect the properties of a vector or its behaviour under transformations.

A vector can be also decomposed with respect to "non-fixed" axes which change their orientation as a function of time or space. For example, a vector in three dimensional space can be decomposed with respect to two axes, respectively *normal*, and *tangent* to a surface. Moreover, the *radial* and *tangential components* of a vector relate to the *radius of rotation* of an object. The former is parallel to the radius and the latter is orthogonal to it.

In these cases, each of the components may be in turn decomposed with respect to a fixed coordinate system or basis set (e.g., a *global* coordinate system, or inertial reference frame).

Basic Properties

The following section uses the Cartesian coordinate system with basis vectors

$$\mathbf{e}_1 = (1,0,0), \mathbf{e}_2 = (0,1,0), \mathbf{e}_3 = (0,0,1)$$

and assumes that all vectors have the origin as a common base point. A vector a will be written as

$$\mathbf{a} = a_1\mathbf{e}_1 + a_2\mathbf{e}_2 + a_3\mathbf{e}_3.$$

Equality

Two vectors are said to be equal if they have the same magnitude and direction. Equivalently they will be equal if their coordinates are equal. So two vectors

$$\mathbf{a} = a_1\mathbf{e}_1 + a_2\mathbf{e}_2 + a_3\mathbf{e}_3$$

and $$\mathbf{b} = b_1\mathbf{e}_1 + b_2\mathbf{e}_2 + b_3\mathbf{e}_3$$

are equal if

$$a_1 = b_1, \quad a_2 = b_2, \quad a_3 = b_3.$$

Addition and subtraction

Assume now that a and b are not necessarily equal vectors, but that they may have different magnitudes and directions. The sum of a and b is

$$\mathbf{a} + \mathbf{b} = (a_1 + b_1)\mathbf{e}_1 + (a_2 + b_2)\mathbf{e}_2 + (a_3 + b_3)\mathbf{e}_3.$$

The addition may be represented graphically by placing the start of the arrow b at the tip of the arrow a, and then drawing an arrow from the start of a to the tip of b. The new arrow drawn represents the vector a + b, as illustrated below:

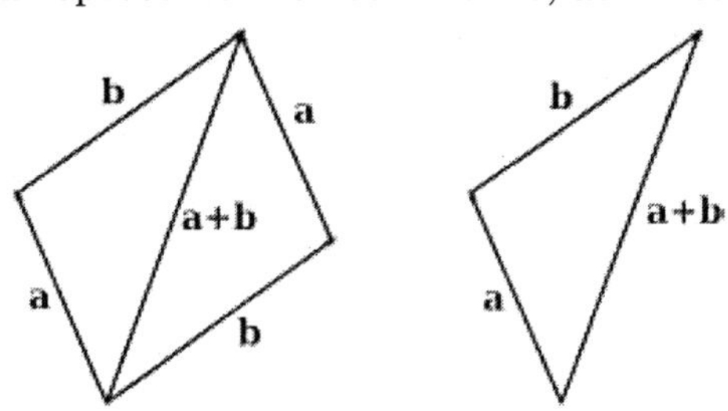

This addition method is sometimes called the *parallelogram rule* because a and b form the sides of a parallelogram and a + b is one of the diagonals. If a and b are bound vectors that have the same base point, it will also be the base point of a + b. One can check geometrically that a + b = b + a and (a + b) + c = a + (b + c).

The difference of a and b is:

$$\mathbf{a} - \mathbf{b} = (a_1 - b_1)\mathbf{e}_1 + (a_2 - b_2)\mathbf{e}_2 + (a_3 - b_3)\mathbf{e}_3.$$

Subtraction of two vectors can be geometrically defined as follows: to subtract b from a, place the end points of a and b at the same point, and then draw an arrow from the tip of b to the tip of a. That arrow represents the vector a “ b, as illustrated below:

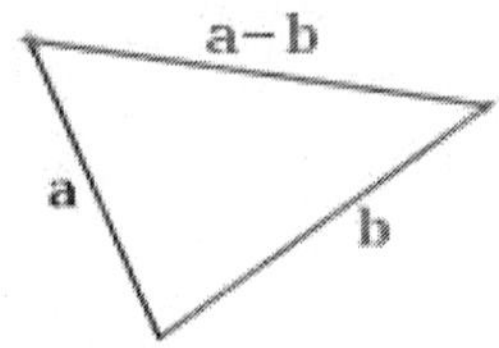

Scalar Multiplication

A vector may also be multiplied, or re-*scaled*, by a real number *r*. In the context of conventional vector algebra, these real numbers are often called scalars (from *scale*) to distinguish them from vectors. The operation of multiplying a vector by a scalar is called *scalar multiplication*. The resulting vector is

$$r\mathbf{a} = (ra_1)\mathbf{e}_1 + (ra_2)\mathbf{e}_2 + (ra_3)\mathbf{e}_3.$$

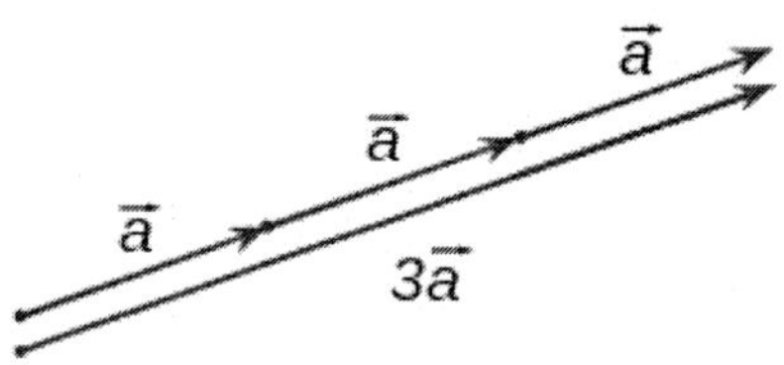

Figure: *Scalar multiplication of a vector by a factor of 3 stretches the vector out.*

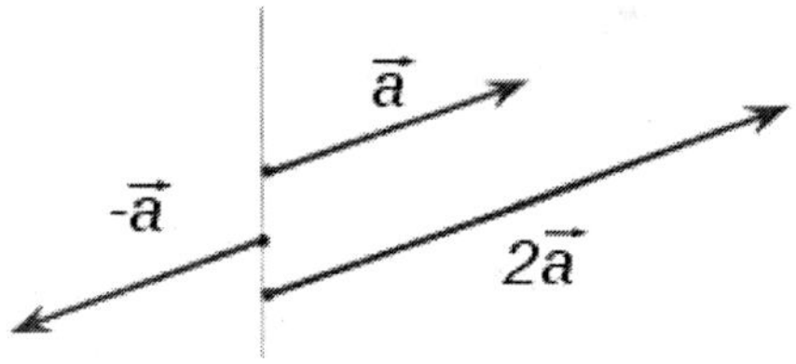

Figure: *The scalar multiplications 2a and -a of a vector a*

Intuitively, multiplying by a scalar *r* stretches a vector out by a factor of *r*. Geometrically, this can be visualized (at least in the case when *r* is an integer) as placing *r* copies of the vector in a line where the endpoint of one vector is the initial point of the next vector. If *r* is negative, then the vector changes direction: it flips around by an angle of 180°. Two examples (*r* = -1 and *r* = 2) are given below:

Scalar multiplication is distributive over vector addition in the following sense: *r*(a + b) = *r*a + *r*b for all vectors a and b and all scalars *r*. One can also show that a - b = a + (-1)b.

Length

The *length* or *magnitude* or *norm* of the vector a is denoted by | | a | | or, less commonly, | a |, which is not to be confused with the absolute value (a scalar "norm").

The length of the vector a can be computed with the Euclidean norm

$$\|\mathbf{a}\| = \sqrt{a_1^2 + a_2^2 + a_3^2}$$

which is a consequence of the Pythagorean theorem since the basis vectors e_1, e_2, e_3 are orthogonal unit vectors.

This happens to be equal to the square root of the dot product, discussed below, of the vector with itself:

$$\|\mathbf{a}\| = \sqrt{\mathbf{a} \cdot \mathbf{a}}.$$

Null Vector

In linear algebra, the null vector or zero vector is the vector (0, 0, ..., 0) in Euclidean space, all of whose components are zero. It is usually written with an arrow head above or below it : $\vec{0}$ or 0 or simply 0. A zero vector has arbitrary direction, but is orthogonal (i.e. perpendicular, normal) to all other vectors with the same number of components.

In vector spaces with an inner product for which the requirement of positive-definiteness has been dropped, a vector that has zero length is referred to as a null vector. The term *zero vector* is then still reserved for the additive identity of the vector spaces.

Linear Algebra

For a general vector space, the zero vector is the uniquely determined vector that is the identity element for vector addition.

The zero vector is unique; if a and b are zero vectors, then a = a + b = b.

The zero vector is a special case of the zero tensor. It is the result of scalar multiplication by the scalar 0 (here meaning the additive identity of the underlying field, not necessarily the real number 0).

The preimage of the zero vector under a linear transformation *f* is called kernel or null space. A zero space is a linear space whose only element is a zero vector. The zero vector is, by itself, linearly dependent, and so any set of vectors which includes it is also linearly dependent.

In a normed vector space there is only one vector of norm equal to 0. This is just the zero vector. In vector algebra its coordinates are (0,0) and its unit vector is n

Seminormed Vector Spaces

In a seminormed vector space there might be more than one vector of norm equal to 0. These vectors are often called null vectors.

Examples

The light-like vectors of Minkowski space are null vectors. In general, the coordinate representation of a null vector in Minkowski space contains non-zero values. In the Verma module of a Lie algebra there are null vectors.

Dot Product

In mathematics, the dot product or scalar product is an algebraic operation that takes two equal-length sequences of numbers (usually coordinate vectors) and returns a single number obtained by multiplying corresponding entries and then summing those products.

The name "dot product" is derived from the centred dot " . " that is often used to designate this operation; the alternative name "scalar product" emphasizes the scalar (rather than vector) nature of the result.

When two Euclidean vectors are expressed in terms of coordinate vectors on an orthonormal basis, the inner product of the former is equal to the dot product of the latter. For more general vector spaces, while both the inner and the dot product can be defined in different contexts (for instance with complex numbers as scalars) their definitions in these contexts may not coincide. In three dimensional space, the dot product contrasts with the cross product, which produces a vector as result. The dot product is directly related to the cosine of the angle between two vectors in Euclidean space of any number of dimensions.

Definition

The dot product of two vectors a = $[a_1, a_2, ..., a_n]$ and b = $[b_1, b_2, ..., b_n]$ is defined as:

$$\mathbf{a}\cdot\mathbf{b} = \sum_{i=1}^{n} a_i b_i = a_1 b_1 + a_2 b_2 + \cdots + a_n b_n$$

where Σ denotes summation notation and n is the dimension of the vector space.

In dimension 2, the dot product of vectors [a,b] and [c,d] is ac + bd. Similarly, in a dimension 3, the dot product of vectors [a,b,c] and [d,e,f] is ad + be + cf. For example, the dot product of two three-dimensional vectors [1, 3, -5] and [4, -2, -1] is

$$[1,3,-5]\cdot[4,-2,-1] = (1)(4) + (3)(-2) + (-5)(-1) = 4 - 6 + 5 = 3.$$

Given two column vectors, their dot product can also be obtained by multiplying the transpose of one vector with the other vector and extracting the unique coefficient of the resulting 1 × 1 matrix. The operation of extracting the coefficient of such a matrix can be written as taking its determinant or its trace (which is the same thing for 1 × 1 matrices); since in general tr(*AB*) = tr(*BA*) whenever *AB* or equivalently *BA* is a square matrix, one may write

$$\mathbf{a}\cdot\mathbf{b} = \det(\mathbf{a}^{\mathrm{T}}\mathbf{b}) = \det(\mathbf{b}^{\mathrm{T}}\mathbf{a}) = \mathrm{tr}(\mathbf{a}^{\mathrm{T}}\mathbf{b}) = \mathrm{tr}(\mathbf{b}^{\mathrm{T}}\mathbf{a}) = \mathrm{tr}(\mathbf{a}\mathbf{b}^{\mathrm{T}}) = \mathrm{tr}(\mathbf{b}\mathbf{a}^{\mathrm{T}})$$

More generally the coefficient (i,j) of a product of matrices is the dot product of the transpose of row i of the first matrix and column j of the second matrix.

Geometric Interpretation

Since $\mathbf{A}\cdot\mathbf{B} = \|\mathbf{A}\|\|\mathbf{B}\|\cos\theta$, then $\mathbf{A}_B = \dfrac{\mathbf{A}\cdot\mathbf{B}}{\|\mathbf{B}\|}$.

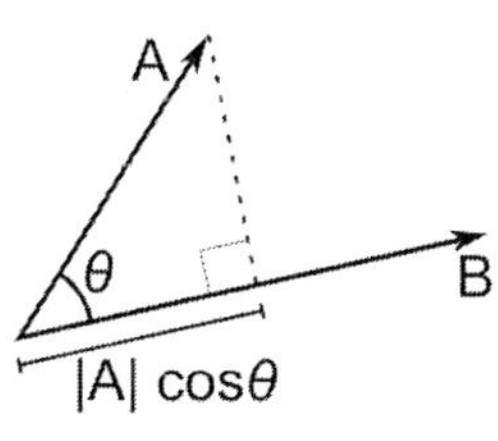

Figure: $\mathbf{A}_B = \|\mathbf{A}\|\cos\theta$ *is the scalar projection of Aonto B.*

In Euclidean geometry, the dot product of vectors expressed in an orthonormal basis is related to their length and angle. For such a vector $\mathbf{a}$, the dot product $\mathbf{a}\cdot\mathbf{a}$ is the square of the length (magnitude) of $\mathbf{a}$, denoted by $\|\mathbf{a}\|$:

$$\mathbf{a}\cdot\mathbf{a} = \|\mathbf{a}\|^2$$

If b is another such vector, and θ is the angle between them:

$$\mathbf{a}\cdot\mathbf{b} = \|\mathbf{a}\|\|\mathbf{b}\|\cos\theta$$

This formula can be rearranged to determine the size of the angle between two nonzero vectors:

$$\theta = \arccos\left(\frac{\mathbf{a}\cdot\mathbf{b}}{\|\mathbf{a}\|\|\mathbf{b}\|}\right).$$

The Cauchy–Schwarz inequality guarantees that the argument of is valid.

One can also first convert the vectors to unit vectors by dividing by their magnitude:

$$\hat{\mathbf{a}} = \frac{\mathbf{a}}{\|\mathbf{a}\|}$$

then the angle θ ισ given by

$$\theta = \arccos(\hat{\mathbf{a}}\cdot\hat{\mathbf{b}}).$$

The terminal points of both unit vectors lie on the unit circle. The unit circle is where the trigonometric values for the six trig functions are found. After substitution, the first vector component is cosine and the second vector component is sine, i.e. $(\cos x, \sin x)$ for some angle x. The dot product of the two unit vectors then takes $(\cos x, \sin x)$ and $(\cos y, \sin y)$ for angles and x and y returns $\cos x \cos y + \sin x \sin y = \cos(x - y)$ where $x - y = \theta$.

As the cosine of 90° is zero, the dot product of two orthogonal vectors is always zero. Moreover, two vectors can be considered orthogonal if and only if their dot product is zero, and they have non-null length. This property provides a simple method to test the condition of orthogonality.

Sometimes these properties are also used for "defining" the *dot product*, especially in 2 and 3 dimensions; this definition is equivalent to the above one. For higher dimensions the formula can be used to define the *concept of angle.*

The geometric properties rely on the basis being orthonormal, i.e. composed of pairwise perpendicular vectors with unit length.

Scalar Projection

If both a and b have length one (i.e., they are unit vectors), their dot product simply gives the cosine of the angle between them.

If only b is a unit vector, then the dot product $\mathbf{a}\cdot\mathbf{b}$ gives $\|\mathbf{a}\|\cos\theta$, i.e., the magnitude of the projection of in the direction of b, with a minus sign if the direction is opposite. This is called the scalar projection of a onto b, or scalar component of in the direction of . This property of the dot product has several useful applications. If neither nor is a unit vector, then the magnitude of the projection of a in the direction of b is $\mathbf{a}\cdot\frac{\mathbf{b}}{\|\mathbf{b}\|}$, as the unit vector a in the direction of b is $\frac{\mathbf{b}}{\|\mathbf{b}\|}$.

Rotation

When an orthonormal basis that the vector is represented in terms of is rotated, 's matrix in the new basis is obtained through multiplying by a rotation matrix . This matrix multiplication is just a compact representation of a sequence of dot products.

For instance, let

- $\mathbf{B}_1 = \{\mathbf{x}, \mathbf{y}, \mathbf{z}\}$ and $\mathbf{B}_2 = \{\mathbf{u}, \mathbf{v}, \mathbf{w}\}$ be two different orthonormal bases of the same space $\mathbb{R}^3$, with $\mathbf{B}_2$ obtained by just rotating $\mathbf{B}_1$,
- $\mathbf{a}_1 = (a_x, a_y, a_z)$ represent vector a in terms of $\mathbf{B}_1$,

- $\mathbf{a}_2 = (a_u, a_v, a_w)$ represent the same vector in terms of the rotated basis $\mathbf{B}_2$,
- $\mathbf{u}_1$, $\mathbf{v}_1$, $\mathbf{w}_1$, be the rotated basis vectors u, v, w represented in terms of $\mathbf{B}_1$.

Then the rotation from $\mathbf{B}_1$ to $\mathbf{B}_2$ is performed as follows:

$$\mathbf{a}_2 = \mathbf{R}\mathbf{a}_1 = \begin{bmatrix} u_x & u_y & u_z \\ v_x & v_y & v_z \\ w_x & w_y & w_z \end{bmatrix} \begin{bmatrix} a_x \\ a_y \\ a_z \end{bmatrix} = \begin{bmatrix} \mathbf{u}_1 \cdot \mathbf{a}_1 \\ \mathbf{v}_1 \cdot \mathbf{a}_1 \\ \mathbf{w}_1 \cdot \mathbf{a}_1 \end{bmatrix} = \begin{bmatrix} a_u \\ a_v \\ a_w \end{bmatrix}.$$

Notice that the rotation matrix R is assembled by using the rotated basis vectors $\mathbf{u}_1, \mathbf{v}_1, \mathbf{w}_1$ as its rows, and these vectors are unit vectors. By definition, $\mathbf{R}\mathbf{a}_1$ consists of a sequence of dot products between each of the three rows of R and vector $\mathbf{a}_1$. Each of these dot products determines a scalar component of a in the direction of a rotated basis vector. If $\mathbf{a}_1$ is a row vector, rather than a column vector, then R must contain the rotated basis vectors in its columns, and must post-multiply $\mathbf{a}_1$:

$$\mathbf{a}_2 = \mathbf{a}_1\mathbf{R} = \begin{bmatrix} a_x & a_y & a_z \end{bmatrix} \begin{bmatrix} u_x & v_x & w_x \\ u_y & v_y & w_y \\ u_z & v_z & w_z \end{bmatrix} = \begin{bmatrix} \mathbf{u}_1 \cdot \mathbf{a}_1 & \mathbf{v}_1 \cdot \mathbf{a}_1 & \mathbf{w}_1 \cdot \mathbf{a}_1 \end{bmatrix} = \begin{bmatrix} a_u & a_v & a_w \end{bmatrix}.$$

Physics

In physics, vector magnitude is a scalar in the physical sense, i.e. a physical quantity independent of the coordinate system, expressed as the product of a numerical value and a physical unit, not just a number.

The dot product is also a scalar in this sense, given by the formula, independent of the coordinate system. Example:

- Mechanical work is the dot product of force and displacement vectors.

- Magnetic flux is the dot product of the magnetic field and the area vectors.

Properties

The following properties hold if a, b, and c are real vectors and r is a scalar.

The dot product is commutative:

$$\mathbf{a} \cdot \mathbf{b} = \mathbf{b} \cdot \mathbf{a}.$$

The dot product is distributive over vector addition:

$$\mathbf{a} \cdot (\mathbf{b} + \mathbf{c}) = \mathbf{a} \cdot \mathbf{b} + \mathbf{a} \cdot \mathbf{c}.$$

The dot product is bilinear:

$$\mathbf{a} \cdot (r\mathbf{b} + \mathbf{c}) = r(\mathbf{a} \cdot \mathbf{b}) + (\mathbf{a} \cdot \mathbf{c}).$$

When multiplied by a scalar value, dot product satisfies:

$$(c_1\mathbf{a}) \cdot (c_2\mathbf{b}) = (c_1 c_2)(\mathbf{a} \cdot \mathbf{b})$$

(these last two properties follow from the first two).

Two non-zero vectors a and b are orthogonal if and only if a • b = 0.

Unlike multiplication of ordinary numbers, where if $ab = ac$, then b always equals c unless a is zero, the dot product does not obey the cancellation law:

If a • b = a • c and a ≠ 0, then we can write: a • (b - c) = 0 by the distributive law; the result above says this just means that a is perpendicular to (b - c), which still allows (b - c) ≠ 0, and therefore b ≠ c. Provided that the basis is orthonormal, the dot product is invariant under isometric changes of the basis: rotations, reflections, and combinations, keeping the origin fixed. The above mentioned geometric interpretation relies on this property. In other words, for an orthonormal space with any number of dimensions, the dot product is invariant under a coordinate transformation based on an orthogonal matrix. This corresponds to the following two conditions:

- The new basis is again orthonormal (i.e., it is orthonormal expressed in the old one).
- The new base vectors have the same length as the old ones (i.e., unit length in terms of the old basis). If a and b are functions, then the derivative of a•b is a' • b + a•b'

Triple Product Expansion

This is a very useful identity (also known as Lagrange's formula) involving the dot- and cross-products. It is written as

$$\mathbf{a} \times (\mathbf{b} \times \mathbf{c}) = \mathbf{b}(\mathbf{a} \cdot \mathbf{c}) - \mathbf{c}(\mathbf{a} \cdot \mathbf{b})$$

which is easier to remember as "BAC minus CAB", keeping in mind which vectors are dotted together. This formula is commonly used to simplify vector calculations in physics.

Proof of the Geometric Interpretation

Consider the element of R^n

$$\mathbf{v} = v_1 \hat{\mathbf{e}}_1 + v_2 \hat{\mathbf{e}}_2 + \ldots + v_n \hat{\mathbf{e}}_n.$$

Repeated application of the Pythagorean theorem yields for its length $|v|$

$$|\mathbf{v}|^2 = v_1^2 + v_2^2 + \ldots + v_n^2.$$

But this is the same as

$$\mathbf{v} \cdot \mathbf{v} = v_1^2 + v_2^2 + \ldots + v_n^2,$$

so we conclude that taking the dot product of a vector v with itself yields the squared length of the vector.

Lemma

$$\mathbf{v} \cdot \mathbf{v} = |\mathbf{v}|^2 .$$

Now consider two vectors a and b extending from the origin, separated by an angle θ. A third vector c may be defined as

$$\mathbf{c} \overset{\text{def}}{=} \mathbf{a} - \mathbf{b}.$$

creating a triangle with sides a, b, and c. According to the law of cosines, we have

$$|\mathbf{c}|^2 = |\mathbf{a}|^2 + |\mathbf{b}|^2 - 2|\mathbf{a}||\mathbf{b}|\cos\theta.$$

Substituting dot products for the squared lengths according to Lemma 1, we get $\mathbf{c} \cdot \mathbf{c} = \mathbf{a} \cdot \mathbf{a} + \mathbf{b} \cdot \mathbf{b} - 2|\mathbf{a}||\mathbf{b}|\cos\theta$.

But as $c \equiv a - b$, we also have $\mathbf{c} \cdot \mathbf{c} = (\mathbf{a} - \mathbf{b}) \cdot (\mathbf{a} - \mathbf{b})$, which, according to the distributive law, expands to

$$\mathbf{c} \cdot \mathbf{c} = \mathbf{a} \cdot \mathbf{a} + \mathbf{b} \cdot \mathbf{b} - 2(\mathbf{a} \cdot \mathbf{b}).$$

Merging the two c • c equations, *(1)* and *(2)*, we obtain

$$\mathbf{a}\cdot\mathbf{a}+\mathbf{b}\cdot\mathbf{b}-2(\mathbf{a}\cdot\mathbf{b})=\mathbf{a}\cdot\mathbf{a}+\mathbf{b}\cdot\mathbf{b}-2\,|\,\mathbf{a}\,||\,\mathbf{b}\,|\cos\theta.$$

Subtracting a • a + b • b from both sides and dividing by "2 leaves

$$\mathbf{a}\cdot\mathbf{b}=|\,\mathbf{a}\,||\,\mathbf{b}\,|\cos\theta.$$

Generalization

Real Vector Spaces

The inner product generalizes the dot product to abstract vector spaces over the real numbers and is usually denoted by $\langle\mathbf{a},\mathbf{b}\rangle$. More precisely, if V is a vector space over R, the inner product is a function $V\times V\rightarrow\mathbb{R}$. Owing to the geometric interpretation of the dot product, the norm | | a | | of a vector a in such an inner product space is defined as

$$\|\mathbf{a}\|=\sqrt{\langle\mathbf{a},\mathbf{a}\rangle}$$

such that it generalizes length, and the angle θ between two vectors a and b by

$$\cos\theta=\frac{\langle\mathbf{a},\mathbf{b}\rangle}{\|\mathbf{a}\|\|\mathbf{b}\|}.$$

In particular, two vectors are considered orthogonal if their inner product is zero

$$\langle\mathbf{a},\mathbf{b}\rangle=0.$$

Complex Vectors

For vectors with complex entries, using the given definition of the dot product would lead to quite different geometric properties. For instance the dot product of a vector with itself can be an arbitrary complex number, and can be zero without the vector being the zero vector; this in turn would have severe consequences for notions like length and angle. Many geometric properties can be salvaged, at the cost of giving up the symmetric and bilinear properties of the scalar product, by alternatively defining

$$\mathbf{a}\cdot\mathbf{b}=\sum a_i\overline{b_i}$$

where b_i is the complex conjugate of b_i. Then the scalar product of any vector with itself is a non-negative real number, and it is nonzero except for the zero vector. However this scalar product is not linear in b (but rather conjugate linear), and the scalar product is not symmetric either, since

$$\mathbf{a}\cdot\mathbf{b} = \overline{\mathbf{b}\cdot\mathbf{a}}.$$

The angle between two complex vectors is then given by

$$\cos\theta = \frac{\mathrm{Re}(\mathbf{a}\cdot\mathbf{b})}{\|\mathbf{a}\|\|\mathbf{b}\|}.$$

This type of scalar product is nevertheless quite useful, and leads to the notions of Hermitian form and of general inner product spaces.

The Frobenius inner product generalizes the dot product to matrices. It is defined as the sum of the products of the corresponding components of two matrices having the same size.

Generalization to Tensors

The dot product between a tensor of order n and a tensor of order m is a tensor of order n+m-2. The dot product is calculated by multiplying and summing across a single index in both tensors. If A and B are two tensors with element representation $A_{ij\ldots}^{k\ell\ldots}$ and $B_{mn\ldots}^{p\ldots i}$ the elements of the dot product $\mathbf{A}\cdot\mathbf{B}$ are given by

$$A_{ij\ldots}^{k\ell\ldots}\cdot B_{mn\ldots}^{p\ldots i} = \sum_{i=1}^{n} A_{ij\ldots}^{k\ell\ldots} B_{mn\ldots}^{p\ldots i}.$$

This definition naturally reduces to the standard vector dot product when applied to vectors, and matrix multiplication when applied to matrices.

Occasionally, a double dot product is used to represent multiplying and summing across two indices. The double dot product between two 2nd order tensors is a scalar quantity.

Cross Product

In mathematics, the cross product, vector product, or Gibbs' vector product is a binary operation on two vectors in three-dimensional space. It results in a vector which is perpendicular to both of the vectors being multiplied and therefore normal to the plane containing them. It has many applications in mathematics, engineering and physics. If either of the vectors being multiplied is zero or the vectors are parallel then their cross product is zero. More generally, the magnitude of the product equals the area of a parallelogram with the vectors for sides; in particular for perpendicular vectors this is a rectangle and the magnitude of the product is the product of their lengths. The cross product is anticommutative, distributive over addition and satisfies the Jacobi identity. The space and product form an algebra over a field, which is neither commutative nor associative, but is a Lie algebra with the cross product being the Lie bracket.

Like the dot product, it depends on the metric of Euclidean space, but unlike the dot product, it also depends on the choice of orientation or "handedness".

The product can be generalized in various ways; it can be made independent of orientation by changing the result to pseudovector, or in arbitrary dimensions the exterior product of vectors can be used with a bivector or two-form result. Also, using the orientation and metric structure just as for the traditional 3-dimensional cross product, one can in n dimensions take the product of $n-1$ vectors to produce a vector perpendicular to all of them. But if the product is limited to non-trivial binary products with vector results, it exists only in three and seven dimensions.

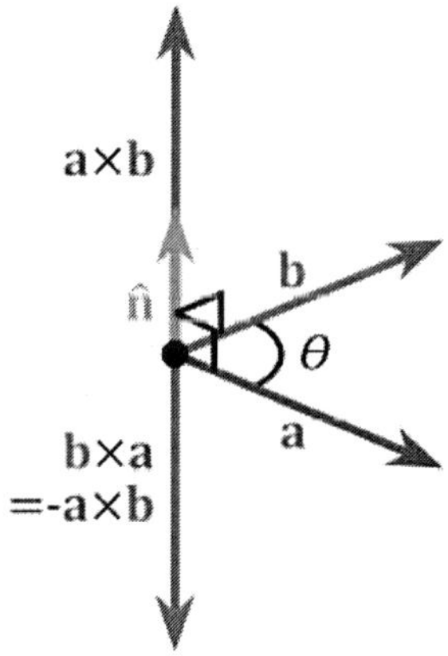

Figure: *The cross-product in respect to a right-handed coordinate system*

Definition

The cross product of two vectors a and b is denoted by a × b. In physics, sometimes the notation a $\wedge$ b is used, though this is avoided in mathematics to avoid confusion with the exterior product.

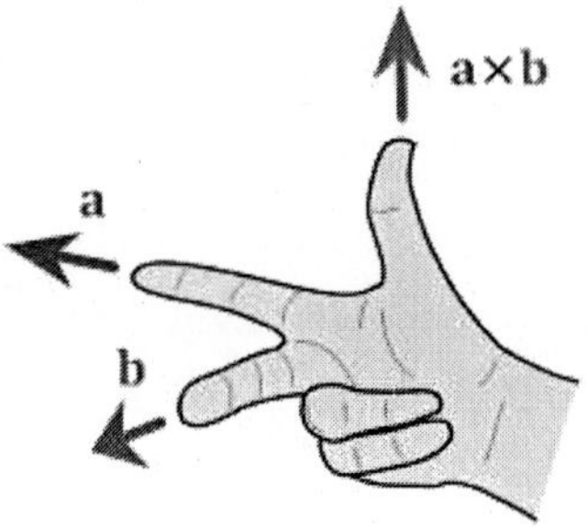

Figure: *Finding the direction of the cross product by the right-hand rule*

The cross product a × b is defined as a vector c that is perpendicular to both a and b, with a direction given by the right-hand rule and a magnitude equal to the area of the parallelogram that the vectors span.

The cross product is defined by the formula

$$\mathbf{a} \times \mathbf{b} = |\mathbf{a}||\mathbf{b}| \sin\theta \, \mathbf{n}$$

where θ is the measure of the smaller angle between a and b ($0° \leq \theta \leq 180°$), |a| and |b| are the magnitudes of vectors a and b, and n is a unit vector perpendicular to the plane containing a and b in the direction given by the right-hand rule as illustrated. If the vectors a and b are parallel (i.e., the angle θ between them is either 0° or 180°), by the above formula, the cross product of a and b is the zero vector 0.

The direction of the vector n is given by the right-hand rule, where one simply points the forefinger of the right hand in the direction of a and the middle finger in the direction of b. Then, the vector n is coming out of the thumb. Using this rule implies that the cross-product is anti-commutative, i.e., b × a = -(a × b). By pointing the forefinger toward b first, and then pointing the middle finger toward a, the thumb will be forced in the opposite direction, reversing the sign of the product vector.

Using the cross product requires the handedness of the coordinate system to be taken into account (as explicit in the definition above). If a left-handed coordinate system is used, the direction of the vector n is given by the left-hand rule and points in the opposite direction.

This, however, creates a problem because transforming from one arbitrary reference system to another (*e.g.*, a mirror image transformation from a right-handed to a left-handed coordinate system), should not change the direction of n. The problem is clarified by realizing that the cross-product of two vectors is not a (true) vector, but rather a *pseudovector*.

Names

The cross product is also called vector product or Gibbs' vector product. The name Gibbs' vector product is after Josiah Willard Gibbs, who around 1881 introduced both the dot product and the cross product, using a dot (a · b) and a cross (a × b) to denote them.

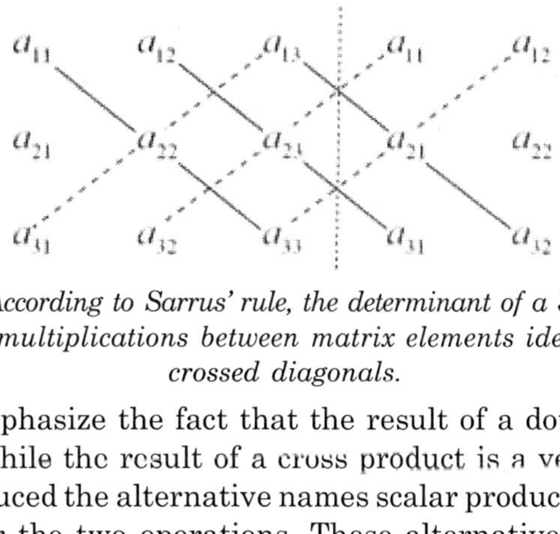

Figure: *According to Sarrus' rule, the determinant of a 3×3 matrix involves multiplications between matrix elements identified by crossed diagonals.*

To emphasize the fact that the result of a dot product is a scalar, while the result of a cross product is a vector, Gibbs also introduced the alternative names scalar product and vector product for the two operations. These alternative names are still widely used in the literature.

Both the cross notation (a × b) and the name cross product were possibly inspired by the fact that each scalar component of a × b is computed by multiplying non-corresponding components of a and b. Conversely, a dot product a · b involves multiplications between corresponding components of a and b.

As explained below, the cross product can be defined as the determinant of a special 3×3 matrix. According to Sarrus' rule, this involves multiplications between matrix elements identified by crossed diagonals.

Computing the Cross Product

Coordinate Notation

The standard basis vectors i, j, and k satisfy the following equalities:

$$\mathbf{i} \times \mathbf{j} = \mathbf{k},$$
$$\mathbf{j} \times \mathbf{k} = \mathbf{i},$$
$$\mathbf{k} \times \mathbf{i} = \mathbf{j},$$

which imply, by the anticommutativity of the cross product, that

$$\mathbf{j} \times \mathbf{i} = -\mathbf{k},$$
$$\mathbf{k} \times \mathbf{j} = -\mathbf{i},$$
$$\mathbf{i} \times \mathbf{k} = -\mathbf{j}.$$

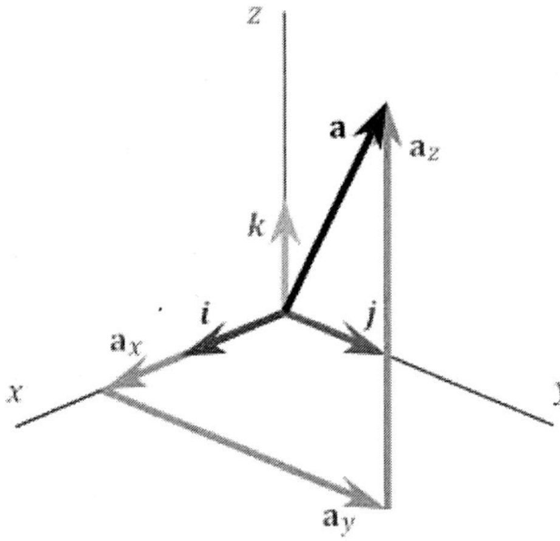

Figure: *Standard basis vectors (i, j, k, also denoted e_1, e_2, e_3) and vector components of a (a_x, a_y, a_z, also denoted a_1, a_2, a_3)*

The definition of the cross product also implies that

$$\mathbf{i} \times \mathbf{i} = \mathbf{j} \times \mathbf{j} = \mathbf{k} \times \mathbf{k} = \mathbf{0} \text{ (the zero vector).}$$

These equalities, together with the distributivity and linearity of the cross product, are sufficient to determine the cross product of any two vectors a and b. Each vector can be defined as the sum of three orthogonal components parallel to the standard basis vectors:

$$\mathbf{a} = \mathbf{a}_1 + \mathbf{a}_2 + \mathbf{a}_3 = a_1\mathbf{i} + a_2\mathbf{j} + a_3\mathbf{k}$$

$$\mathbf{b} = \mathbf{b}_1 + \mathbf{b}_2 + \mathbf{b}_3 = b_1\mathbf{i} + b_2\mathbf{j} + b_3\mathbf{k}$$

Their cross product a × b can be expanded using distributivity:

$$\begin{aligned}\mathbf{a}\times\mathbf{b} &= (a_1\mathbf{i} + a_2\mathbf{j} + a_3\mathbf{k}) \times (b_1\mathbf{i} + b_2\mathbf{j} + b_3\mathbf{k}) \\ &= a_1b_1\mathbf{i}\times\mathbf{i} + a_1b_2\mathbf{i}\times\mathbf{j} + a_1b_3\mathbf{i}\times\mathbf{k} + \\ &\quad a_2b_1\mathbf{j}\times\mathbf{i} + a_2b_2\mathbf{j}\times\mathbf{j} + a_2b_3\mathbf{j}\times\mathbf{k} + \\ &\quad a_3b_1\mathbf{k}\times\mathbf{i} + a_3b_2\mathbf{k}\times\mathbf{j} + a_3b_3\mathbf{k}\times\mathbf{k}\end{aligned}$$

This can be interpreted as the decomposition of a × b into the sum of nine simpler cross products involving vectors aligned with i, j, or k. Each one of these nine cross products operates on two vectors that are easy to handle as they are either parallel or orthogonal to each other. From this decomposition, by using the above mentioned equalities and collecting similar terms, we obtain:

$$\begin{aligned}a\times b = a_1b_1\mathbf{0} + a_1b_2\mathbf{k} + a_1b_3(-\mathbf{j}) + a_2b_1(-\mathbf{k}) + a_2b_2\mathbf{0} + a_2b_3\mathbf{i} \\ + a_3b_1\mathbf{j} + a_3b_2(-\mathbf{i}) + a_3b_3\mathbf{0}\end{aligned}$$

$$= (a_2b_3 - a_3b_2)\mathbf{i} + (a_3b_1 - a_1b_3)\mathbf{j} + (a_1b_2 - a_2b_1)\mathbf{k}.$$

meaning that the three scalar components of the resulting vectorc = c_1i + c_2j + c_3k = a × b are

$$c_1 = a_2b_3 - a_3b_2$$

$$c_2 = a_3b_1 - a_1b_3$$

$$c_3 = a_1b_2 - a_2b_1$$

Using column vectors, we can represent the same result as follows:

$$\begin{pmatrix} c_1 \\ c_2 \\ c_3 \end{pmatrix} = \begin{pmatrix} a_2b_3 - a_3b_2 \\ a_3b_1 - a_1b_3 \\ a_1b_2 - a_2b_1 \end{pmatrix}.$$

Matrix Notation

The definition of the cross product can also be represented by the determinant of a formal matrix:

$$\mathbf{a} \times \mathbf{b} = \begin{vmatrix} \mathbf{i} & \mathbf{j} & \mathbf{k} \\ a_1 & a_2 & a_3 \\ b_1 & b_2 & b_3 \end{vmatrix}.$$

This determinant can be computed using Sarrus' rule or Cofactor expansion.

Using Sarrus' Rule, it expands to

$$\mathbf{a} \times \mathbf{b} = \mathbf{i}a_2b_3 + \mathbf{j}a_3b_1 + \mathbf{k}a_1b_2 - \mathbf{i}a_3b_2 - \mathbf{j}a_1b_3 - \mathbf{k}a_2b_1.$$

Using Cofactor expansion along the first row instead, it expands to

$$\mathbf{a} \times \mathbf{b} = \begin{vmatrix} a_2 & a_3 \\ b_2 & b_3 \end{vmatrix} \mathbf{i} - \begin{vmatrix} a_1 & a_3 \\ b_1 & b_3 \end{vmatrix} \mathbf{j} + \begin{vmatrix} a_1 & a_2 \\ b_1 & b_2 \end{vmatrix} \mathbf{k}$$

which gives the components of the resulting vector directly.

Properties

Geometric meaning: The magnitude of the cross product can be interpreted as the positive area of the parallelogram having a and b as sides:

$$A = |\mathbf{a} \times \mathbf{b}| = |\mathbf{a}||\mathbf{b}| \sin\theta.$$

Indeed, one can also compute the volume V of a parallelepiped having a, b and c as sides by using a combination of a cross product and a dot product, called scalar triple product:

$$\mathbf{a} \cdot (\mathbf{b} \times \mathbf{c}) = \mathbf{b} \cdot (\mathbf{c} \times \mathbf{a}) = \mathbf{c} \cdot (\mathbf{a} \times \mathbf{b}).$$

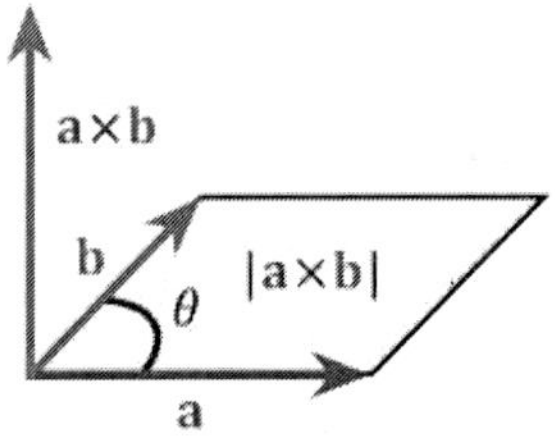

Figure: *The area of a parallelogram as a cross product*

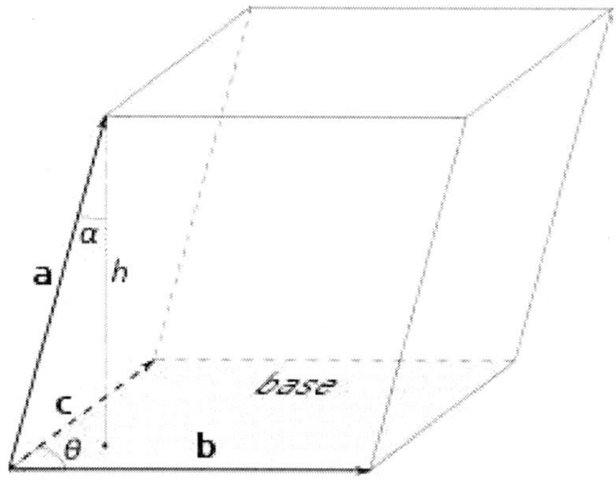

Figure: *Three vectors defining a parallelepiped*

Since the result of the scalar triple product may be negative, the volume of the parallelepiped is given by its absolute value. For instance,

$$V = |\mathbf{a}\cdot(\mathbf{b}\times\mathbf{c})|.$$

Because the magnitude of the cross product goes by the sine of the angle between its arguments, the cross product can be thought of as a measure of "perpendicularness" in the same way that the dot product is a measure of "parallelness". Given two unit vectors, their cross product has a magnitude of 1 if the two are perpendicular and a magnitude of zero if the two are parallel. The opposite is true for the dot product of two unit vectors.

Unit vectors enable two convenient identities: the dot product of two unit vectors yields the cosine (which may be positive or negative) of the angle between the two unit vectors. The magnitude of the cross product of the two unit vectors yields the sine (which will always be positive).

Algebraic Properties

The cross product is anticommutative,

$$\mathbf{a}\times\mathbf{b} = -\mathbf{b}\times\mathbf{a},$$

distributive over addition,

$$\mathbf{a}\times(\mathbf{b}+\mathbf{c}) = (\mathbf{a}\times\mathbf{b})+(\mathbf{a}\times\mathbf{c}),$$

and compatible with scalar multiplication so that

$$(r\mathbf{a})\times\mathbf{b} = \mathbf{a}\times(r\mathbf{b}) = r(\mathbf{a}\times\mathbf{b}).$$

It is not associative, but satisfies the Jacobi identity:

$$\mathbf{a}\times(\mathbf{b}\times\mathbf{c})+\mathbf{b}\times(\mathbf{c}\times\mathbf{a})+\mathbf{c}\times(\mathbf{a}\times\mathbf{b})=\mathbf{0}.$$

Distributivity, linearity and Jacobi identity show that R^3 together with vector addition and the cross product forms a Lie algebra, the Lie algebra of the real orthogonal group in 3 dimensions, SO(3).

The cross product does not obey the cancellation law: a × b = a × c with non-zero a does not imply that b = c. Instead if a × b = a × c:

$$\begin{aligned}\mathbf{0} &= (\mathbf{a}\times\mathbf{b})-(\mathbf{a}\times\mathbf{c})\\ &= \mathbf{a}\times(\mathbf{b}-\mathbf{c}).\end{aligned}$$

If neither a nor b - c is zero then from the definition of the cross product the angle between them must be zero and they must be parallel. They are related by a scale factor, so one of b or c can be expressed in terms of the other, for example

$$\mathbf{c}=\mathbf{b}+t\mathbf{a},$$

for some scalar t.

If a · b = a · c and a × b = a × c, for non-zero vector a, then b = c, as

$$\mathbf{a}\times(\mathbf{b}-\mathbf{c})=\mathbf{0}$$

and $$\mathbf{a}\cdot(\mathbf{b}-\mathbf{c})=0,$$

So b - c is both parallel and perpendicular to the non-zero vector a, something that is only possible if b - c = 0 so they are identical. From the geometrical definition the cross product is invariant under rotations about the axis defined by a × b. More generally the cross product obeys the following identity under matrix transformations:

$$(M\mathbf{a})\times(M\mathbf{b})=(\det M)M^{-T}(\mathbf{a}\times\mathbf{b})$$

where is a 3 by 3 matrix and M^{-T} is the transpose of the inverse

The cross product of two vectors in 3-D always lies in the null space of the matrix with the vectors as rows:

$$\mathbf{a}\times\mathbf{b}\in NS\left(\begin{bmatrix}\mathbf{a}\\ \mathbf{b}\end{bmatrix}\right).$$

For the sum of two cross products, the following identity holds:

$$\mathbf{a}\times\mathbf{b}+\mathbf{c}\times\mathbf{d}=(\mathbf{a}-\mathbf{c})\times(\mathbf{b}-\mathbf{d})+\mathbf{a}\times\mathbf{d}+\mathbf{c}\times\mathbf{b}.$$

Differentiation

The product rule applies to the cross product in a similar manner:

$$\frac{d}{dx}(\mathbf{a}\times\mathbf{b})=\frac{d\mathbf{a}}{dx}\times\mathbf{b}+\mathbf{a}\times\frac{d\mathbf{b}}{dx}.$$

This identity can be easily proved using the matrix multiplication representation.

Triple Product Expansion

The cross product is used in both forms of the triple product. The scalar triple product of three vectors is defined as

$$\mathbf{a}\cdot(\mathbf{b}\times\mathbf{c}),$$

It is the signed volume of the parallelepiped with edges a, b and c and as such the vectors can be used in any order that's an even permutation of the above ordering. The following therefore are equal:

$$\mathbf{a}\cdot(\mathbf{b}\times\mathbf{c})=\mathbf{b}\cdot(\mathbf{c}\times\mathbf{a})=\mathbf{c}\cdot(\mathbf{a}\times\mathbf{b}),$$

The vector triple product is the cross product of a vector with the result of another cross product, and is related to the dot product by the following formula

$$\mathbf{a}\times(\mathbf{b}\times\mathbf{c})=\mathbf{b}(\mathbf{a}\cdot\mathbf{c})-\mathbf{c}(\mathbf{a}\cdot\mathbf{b}).$$

The mnemonic "BAC minus CAB" is used to remember the order of the vectors in the right hand member. This formula is used in physics to simplify vector calculations. A special case, regarding gradients and useful in vector calculus, is

$$\begin{aligned}\nabla\times(\nabla\times\mathbf{f}) &= \nabla(\nabla\cdot\mathbf{f})-(\nabla\cdot\nabla)\mathbf{f}\\ &= \nabla(\nabla\cdot\mathbf{f})-\nabla^2\mathbf{f},\end{aligned}$$

where ∇^2 is the vector Laplacian operator.

Another identity relates the cross product to the scalar triple product:

$$(\mathbf{a}\times\mathbf{b})\times(\mathbf{a}\times\mathbf{c})=(\mathbf{a}\cdot(\mathbf{b}\times\mathbf{c}))\mathbf{a}$$

Alternative Formulation

The cross product and the dot product are related by:

$$|\mathbf{a}\times\mathbf{b}|^2=|\mathbf{a}|^2|\mathbf{b}|^2-(\mathbf{a}\cdot\mathbf{b})^2.$$

The right-hand side is the Gram determinant of a and b, the square of the area of the parallelogram defined by the vectors. This condition determines the magnitude of the cross product. Namely, since the dot product is defined, in terms of the angle θ between the two vectors, as:

$$\mathbf{a}\cdot\mathbf{b}=|\mathbf{a}||\mathbf{b}|\cos\theta,$$

the above given relationship can be rewritten as follows:

$$|\mathbf{a}\times\mathbf{b}|^2=|\mathbf{a}|^2|\mathbf{b}|^2\left(1-\cos^2\theta\right).$$

Invoking the Pythagorean trigonometric identity one obtains:

$$|\mathbf{a}\times\mathbf{b}|=|\mathbf{a}||\mathbf{b}|\sin\theta,$$

which is the magnitude of the cross product expressed in terms of θ, equal to the area of the parallelogram defined by a and b.

The combination of this requirement and the property that the cross product be orthogonal to its constituents a and b provides an alternative definition of the cross product.

Lagrange's Identity

The relation:

$$|\mathbf{a}\times\mathbf{b}|^2=|\mathbf{a}|^2|\mathbf{b}|^2-(\mathbf{a}\cdot\mathbf{b})^2.$$

can be compared with another relation involving the right-hand side, namely Lagrange's identity expressed as:

$$\sum_{1\le i<j\le n}\left(a_ib_j-a_jb_i\right)^2=|\mathbf{a}|^2|\mathbf{b}|^2-(\mathbf{a}\cdot\mathbf{b})^2,$$

where a and b may be n-dimensional vectors. In the case $n=3$, combining these two equations results in the expression for the magnitude of the cross product in terms of its components:

$$|\mathbf{a}\times\mathbf{b}|^2=\sum_{1\le i<j\le 3}\left(a_ib_j-a_jb_i\right)^2=(a_1b_2-b_1a_2)^2+(a_2b_3-a_3b_2)^2+(a_3b_1-a_1b_3)^2.$$

The same result is found directly using the components of the cross-product found from:

$$\mathbf{a}\times\mathbf{b}=\det\begin{bmatrix}\mathbf{i} & \mathbf{j} & \mathbf{k}\\ a_1 & a_2 & a_3\\ b_1 & b_2 & b_3\end{bmatrix}.$$

In R^3 Lagrange's equation is a special case of the multiplicativity $|vw| = |v||w|$ of the norm in the quaternion algebra. It is a special case of another formula, also sometimes called Lagrange's identity, which is the three dimensional case of the Binet-Cauchy identity:

$$(\mathbf{a}\times\mathbf{b})\cdot(\mathbf{c}\times\mathbf{d})=(\mathbf{a}\cdot\mathbf{c})(\mathbf{b}\cdot\mathbf{d})-(\mathbf{a}\cdot\mathbf{d})(\mathbf{b}\cdot\mathbf{c}).$$

If a = c and b = d this simplifies to the formula above.

Alternative Ways to Compute the Cross Product

Conversion to matrix multiplication:

The vector cross product also can be expressed as the product of a skew-symmetric matrix and a vector:

$$\mathbf{a}\times\mathbf{b}=[\mathbf{a}]_\times\mathbf{b}=\begin{bmatrix}0 & -a_3 & a_2\\ a_3 & 0 & -a_1\\ -a_2 & a_1 & 0\end{bmatrix}\begin{bmatrix}b_1\\ b_2\\ b_3\end{bmatrix}$$

$$\mathbf{a}\times\mathbf{b}=[\mathbf{b}]_\times^{\mathrm{T}}\mathbf{a}=\begin{bmatrix}0 & b_3 & -b_2\\ -b_3 & 0 & b_1\\ b_2 & -b_1 & 0\end{bmatrix}\begin{bmatrix}a_1\\ a_2\\ a_3\end{bmatrix}$$

where superscript ^{T} refers to the transpose operation, and $[a]_\times$ is defined by:

$$[\mathbf{a}]_\times\stackrel{\mathrm{def}}{=}\begin{bmatrix}0 & -a_3 & a_2\\ a_3 & 0 & -a_1\\ -a_2 & a_1 & 0\end{bmatrix}.$$

Also, if a is itself a cross product:

$$\mathbf{a}=\mathbf{c}\times\mathbf{d}$$

then

$$[\mathbf{a}]_\times=(\mathbf{c}\mathbf{d}^{\mathrm{T}})^{\mathrm{T}}-\mathbf{c}\mathbf{d}^{\mathrm{T}}.$$

This result can be generalized to higher dimensions using geometric algebra. In particular in any dimension bivectors can be identified with skew-symmetric matrices, so the product between a skew-symmetric matrix and vector is equivalent to the grade-1 part of the product of a bivector and vector. In three dimensions bivectors are dual to vectors so the product is equivalent to the cross product, with the bivector instead of its vector dual. In higher dimensions the product can still be calculated but bivectors have more degrees of freedom and are not equivalent to vectors.

This notation is also often much easier to work with, for example, in epipolar geometry.

From the general properties of the cross product follows immediately that

$$[\mathbf{a}]_{\times}\mathbf{a} = \mathbf{0} \quad \text{and} \quad \mathbf{a}^{\mathrm{T}}[\mathbf{a}]_{\times} = \mathbf{0}$$

and from fact that $[a]_{\times}$ is skew-symmetric it follows that

$$\mathbf{b}^{\mathrm{T}}[\mathbf{a}]_{\times}\mathbf{b} = 0.$$

The above-mentioned triple product expansion (bac-cab rule) can be easily proven using this notation.

The above definition of $[a]_{\times}$ means that there is a one-to-one mapping between the set of 3×3 skew-symmetric matrices, also known as the Lie algebra of SO(3), and the operation of taking the cross product with some vector a.

Index Notation for Tensors

The cross product can alternatively be defined in terms of the Levi-Civita symbol, å$_{\mathrm{ijk}}$ which is useful in converting vector notation for tensor applications:

$$\mathbf{a} \times \mathbf{b} = \mathbf{c} \Leftrightarrow c_i = \sum_{j=1}^{3}\sum_{k=1}^{3} \varepsilon_{ijk} a_j b_k$$

where the indices i, j, k correspond, as in the previous section, to orthogonal vector components. This characterization of the cross product is often expressed more compactly using the Einstein summation convention as

$$\mathbf{a} \times \mathbf{b} = \mathbf{c} \Leftrightarrow c_i = \varepsilon_{ijk} a_j b_k$$

in which repeated indices are summed from 1 to 3. Note that this representation is another form of the skew-symmetric representation of the cross product:

$$\varepsilon_{ijk} a_j = [\mathbf{a}]_\times .$$

In Classical Mechanics: representing the cross-product with the Levi-Civita symbol can cause mechanical-symmetries to be obvious when physical-systems are isotropic in space. (Quick example: consider a particle in a Hooke's Law potential in three-space, free to oscillate in three dimensions; none of these dimensions are "special" in any sense, so symmetries lie in the cross-product-represented angular-momentum which are made clear by the above mentioned Levi-Civita representation).

Mnemonic

The word "xyzzy" can be used to remember the definition of the cross product.

If $\mathbf{a} = \mathbf{b} \times \mathbf{c}$

where:

$$\mathbf{a} = \begin{bmatrix} a_x \\ a_y \\ a_z \end{bmatrix}, \mathbf{b} = \begin{bmatrix} b_x \\ b_y \\ b_z \end{bmatrix}, \mathbf{c} = \begin{bmatrix} c_x \\ c_y \\ c_z \end{bmatrix}$$

then:

$$a_x = b_y c_z - b_z c_y$$

$$a_y = b_z c_x - b_x c_z$$

$$a_z = b_x c_y - b_y c_x .$$

The second and third equations can be obtained from the first by simply vertically rotating the subscripts, $x \to y \to z \to x$. The problem, of course, is how to remember the first equation, and two options are available for this purpose: either to remember the relevant two diagonals of Sarrus's scheme (those containing *i*), or to remember the xyzzy sequence.

Since the first diagonal in Sarrus's scheme is just the main diagonal of the above-mentioned matrix, the first three letters of the word xyzzy can be very easily remembered.

Cross Visualization

Similarly to the mnemonic device above, a "cross" or X can be visualized between the two vectors in the equation. This may help you to remember the correct cross product formula.

If $\mathbf{a} = \mathbf{b} \times \mathbf{c}$

then:

$$\mathbf{a} = \begin{bmatrix} b_x \\ b_y \\ b_z \end{bmatrix} \times \begin{bmatrix} c_x \\ c_y \\ c_z \end{bmatrix}.$$

If we want to obtain the formula for a_x we simply drop the b_x and c_x from the formula, and take the next two components down -

$$a_x = \begin{bmatrix} b_y \\ b_z \end{bmatrix} \times \begin{bmatrix} c_y \\ c_z \end{bmatrix}.$$

It should be noted that when doing this for a_y the next two elements down should "wrap around" the matrix so that after the z component comes the x component. For clarity, when performing this operation for a_y, the next two components should be z and x (in that order). While for a_z the next two components should be taken as x and y.

$$a_y = \begin{bmatrix} b_z \\ b_x \end{bmatrix} \times \begin{bmatrix} c_z \\ c_x \end{bmatrix}, a_z = \begin{bmatrix} b_x \\ b_y \end{bmatrix} \times \begin{bmatrix} c_x \\ c_y \end{bmatrix}$$

For a_x then, if we visualize the cross operator as pointing from an element on the left to an element on the right, we can take the first element on the left and simply multiply by the element that the cross points to in the right hand matrix. We then subtract the next element down on the left, multiplied by the element that the cross points to here as well. This results in our a_x formula-

$$a_x = b_y c_z - b_z c_y.$$

We can do this in the same way for a_y and a_z to construct their associated formulas.

Applications

Computational Geometry: The cross product can be used to calculate the normal for a triangle or polygon, an operation frequently performed in computer graphics. For example, the winding of polygon (clockwise or anticlockwise) about a point within the polygon (i.e. the centroid or mid-point) can be calculated by triangulating the polygon (like spoking a wheel) and summing the angles (between the spokes) using the cross product to keep track of the sign of each angle.

In computational geometry of the plane, the cross product is used to determine the sign of the acute angle defined by three points $p_1 = (x_1, y_1)$, $p_2 = (x_2, y_2)$.

It corresponds to the direction of the cross product of the two coplanar vectors defined by the pairs of points p_1, p_2 and p_1, p_3, i.e., by the sign of the expression $P = (x_2 - x_1)(y_3 - y_1) - (y_2 - y_1)(x_3 - x_1)$. In the "right-handed" coordinate system, if the result is 0, the points are collinear; if it is positive, the three points constitute a negative angle of rotation around from to , otherwise a positive angle. From another point of view, the sign of tells whether lies to the left or to the right of line p_1, p_2.

Chapter 6

Continuity Equation

A continuity equation in physics is an equation that describes the transport of a conserved quantity. Since mass, energy, momentum, electric charge and other natural quantities are conserved under their respective appropriate conditions, a variety of physical phenomena may be described using continuity equations.

Continuity equations are the (stronger) local form of conservation laws. All the examples of continuity equations below express the same idea, which is roughly that: *the total amount (of the conserved quantity) inside any region can only change by the amount that passes in or out of the region through the boundary*. A conserved quantity cannot increase or decrease, it can only move from place to place.

Any continuity equation can be expressed in an "integral form" (in terms of a flux integral), which applies to any finite region, or in a "differential form" (in terms of the divergence operator) which applies at a point. Continuity equations underlie more specific transport equations such as the convection–diffusion equation, Boltzmann transport equation, and Navier-Stokes equations.

General Equation

As stated above, the idea behind the continuity equation is the flow of some property, such as mass, energy, electric charge, momentum, and even probability, through surfaces from one region of space to another. The surfaces, in general, may either be open or closed, real or imaginary, and have an arbitrary shape, but are fixed for the calculation (i.e. not time-varying, which is appropriate since this complicates the maths for no advantage).

Let this property be represented by just one scalar variable, q, and let the volume density of this property (the amount of q per unit volume V) be φ, and the all surfaces be denoted by S. Mathematically, φ is a ratio of two infinitesimal quantities:

$$\varphi = \frac{dq}{dV},$$

which has the dimension [quantity][L]$^{-3}$ (where L is length).

There are different ways to conceive the continuity equation:

1. either the flow of particles carrying the quantity q, described by a velocity field v, which is also equivalent to a flux f of q (a vector function describing the flow per unit area per unit time of q), or
2. in the cases where a velocity field is not useful or applicable, the flux f of the quantity q only (no association with velocity).

In each of these cases, the transfer of q occurs as it passes through two surfaces, the first S_1 and the second S_2.

The flux f should represent some flow or transport, which has dimensions [quantity][T]$^{-1}$[L]$^{-2}$. In cases where particles/ carriers of quantity q are moving with velocity v, such as particles of mass in a fluid or charge carriers in a conductor, f can be related to v by:

$$f = \varphi v.$$

This relation is only true in situations where there are particles moving and carrying q - it can't always be applied. To illustrate this: if f is electric current density (electric current per unit area) and φ is the charge density (charge per unit volume), then the velocity of the charge carriers is v.

However - if f is heat flux density (heat energy per unit time per unit area), then even if we let φ be the heat energy density (heat energy per unit volume) it does *not* imply the "velocity of heat" is v (this makes no sense, and is not practically applicable). In the latter case only f (with φ) may be used in the continuity equation.

Elementary Vector Form

Consider the case when the surfaces are flat and planar cross-sections. For the case where a velocity field can be applied, dimensional analysis leads to this form of the continuity equation:

$$\varphi_1 v_1 \cdot S_1 = \varphi_2 v_2 \cdot S_2$$

where

- the left hand side is the initial amount of q flowing per unit time through surface S_1, the right hand side is the final amount through surface S_2,
- S_1 and S_2 are the vector areas for the surfaces S_1 and S_2 respectively.

Notice the dot products $v_1 \cdot S_1, v_2 \cdot S_2$ are volumetric flow rates of q. The dimension of each side of the equation is $[\text{quantity}][L]^{-3} \cdot [L][T]^{-1} \cdot [L]^2 = [\text{quantity}][T]^{-1}$. For the more general cases, independent of whether a velocity field can be used or not, the continuity equation becomes:

$$f_1 \cdot S_1 = f_2 \cdot S_2$$

This has exactly the same dimensions as the previous version. The relation between f and v allows us to pass back to the velocity version from this flux equation, but *not* always the other way round (as explained above - velocity fields are not always applicable). These results can be generalized further to curved surfaces by reducing the vector surfaces into infinitely many differential surface elements (that is $\mathrm{S} \to \mathrm{dS}$), then integrating over the surface:

$$\iint_{S_1} \varphi_1 v_1 \cdot \mathrm{d}S_1 = \iint_{S_2} \varphi_2 v_2 \cdot \mathrm{d}S_2$$

more generally still:

$$\iint_{S_1} f_1 \cdot \mathrm{d}S_1 = \iint_{S_2} f_2 \cdot \mathrm{d}S_2$$

in which

- $\iint_S \mathrm{d}S \equiv \iint_S \hat{n} \mathrm{d}S$ denotes a surface integral over the surface S,
- $\hat{n}$ is the outward-pointing unit normal to the surface S

N.B: the scalar area S and vector area S are related by $\mathrm{d}S = \hat{n} \mathrm{d}S$. Either notations may be used interchangeably.

Differential Form

The differential form for a general continuity equation is (using the same q, φ and f as above):

$$\frac{\partial \varphi}{\partial t} + \nabla \cdot f = \sigma$$

where

- ∇
- is divergence,
- t is time,
- σ is the generation of q per unit volume per unit time. Terms that generate ($\Sigma > 0$) or remove ($\sigma < 0$) q are referred to as a "sources" and "sinks" respectively.

This general equation may be used to derive any continuity equation, ranging from as simple as the volume continuity equation to as complicated as the Navier–Stokes equations. This equation also generalizes the advection equation. Other equations in physics, such as Gauss's law of the electric field and Gauss's law for gravity, have a similar mathematical form to the continuity equation, but are not usually called by the term "continuity equation", because f in those cases does not represent the flow of a real physical quantity.

In the case that q is a conserved quantity that cannot be created or destroyed (such as energy), this translates to $\sigma = 0$, and the continuity equation is:

$$\frac{\partial \varphi}{\partial t} + \nabla \cdot f = 0$$

Integral Form

By the divergence theorem, the continuity equation can be rewritten in an equivalent way, called the "integral form":

$$\frac{\mathrm{d}q}{\mathrm{d}t} + \oiint s f \cdot \mathrm{d}S = \Sigma$$

where

- S is a surface as described above - except this time it has to be a closed surface that encloses a volume V,
- $\oiint s dS$ denotes a surface integral over a closed surface,

- $\iiint_V dV$ denotes a volume integral over V.
- $q = \iiint_V \varphi dV$ is the total amount of φ in the volume V;
- $\Sigma = \iiint_V \sigma dV$ is the total generation (negative in the case of removal) per unit time by the sources and sinks in the volume V,

In a simple example, V could be a building, and q could be the number of people in the building. The surface S would consist of the walls, doors, roof, and foundation of the building. Then the continuity equation states that the number of people in the building increases when people enter the building (an inward flux through the surface), decreases when people exit the building (an outward flux through the surface), increases when someone in the building gives birth (a "source" where $\sigma > 0$), and decreases when someone in the building dies (a "sink" where $\sigma < 0$).

Derivation and Equivalence

The differential form can be derived from first principles as follows.

Derivation of the Differential Form

Suppose first an amount of quantity q *is contained in a region of volume* V, *bounded by a closed surface* S, *as described above. This is equal to the amount already in* V, *plus the generated amount* s *(total - not per unit time or volume):*

$$q(t) = \iiint_V \varphi(r,t) dV + \int^t \Sigma(t') dt'.$$

(t' is just a dummy variable of the generation integral). The rate of change of q leaving the region is simply the time derivative:

$$\frac{\partial q(t)}{\partial t} = -\frac{\partial}{\partial t} \iiint_V \varphi(r,t) dV + \Sigma(t)$$

where the minus sign has been inserted since the amount of q is *decreasing* in the region. (Partial derivatives are used since they enter the integrand, which is not only a function of time, but also space due to the density nature of φ - differentiation needs only to be with respect to t). The rate of change of q crossing the boundary and leaving the region is:

$$\frac{\partial q(t)}{\partial t} = \oiint s f(r,t) \cdot \mathrm{d}S,$$

so equating these expressions:

$$\oiint s f(r,t) \cdot \mathrm{d}S = -\iiint_V \frac{\partial \varphi(r,t)}{\partial t} \mathrm{d}V + \Sigma(t),$$

Using the divergence theorem on the left-hand side:

$$\iiint_V \nabla \cdot f(r,t) \mathrm{d}V = -\iiint_V \frac{\partial \varphi(r,t)}{\partial t} \mathrm{d}V + \iiint_V \sigma(r,t) \mathrm{d}V.$$

This is only true if the integrands are equal, which directly leads to the differential continuity equation:

$$\nabla \cdot f(r,t) = -\frac{\partial \varphi(r,t)}{\partial t} + \sigma(r,t),$$

$$\nabla \cdot f + \frac{\partial \varphi}{\partial t} = \sigma \rightleftharpoons \nabla \cdot (\varphi v) + \frac{\partial \varphi}{\partial t} = \sigma.$$

Either form may be useful and quoted, both can appear in hydrodynamics and electromagnetism, but for quantum mechanics and energy conservation, only the first may be used. Therefore the first is more general.

Equivalence Between Differential and Integral Form

Starting from the differential form which is for unit volume, multiplying throughout by the infinitesimal volume element dV *and integrating over the region gives the total amounts quantities in the volume of the region (per unit time):*

$$\iiint_V \frac{\partial \varphi(r,t)}{\partial t} \mathrm{d}V + \iiint_V \nabla \cdot f(r,t) \mathrm{d}V = \iiint_V \sigma(r,t) \mathrm{d}V$$

$$\frac{\mathrm{d}}{\mathrm{d}t} \iiint_V \varphi(r,t) \mathrm{d}V + \iiint_V \nabla \cdot f(r,t) \mathrm{d}V = \Sigma(t)$$

again using the fact that V is constant in shape for the calculation, so it is independent of time and the time derivatives can be freely moved out of that integral, ordinary derivatives replace partial derivatives since the integral becomes a function of time only (the integral is evaluated over the region - so the spatial variables become removed from the final expression and t remains the only variable). Using the divergence theorem on the left side

$$\frac{dq(t)}{dt} + \oiint sf(r,t)\cdot dS = \Sigma(t)$$

which is the integral form.

Equivalence Between Elementary and Integral Form

Starting from

$$\iint_{S_1} f_1(r,t)\cdot dS_1 = \iint_{S_2} f_2(r,t)\cdot dS_2$$

the surfaces are equal (since there is only one closed surface), so $S_1 = S_2 = S$ and we can write:

$$\iint_S f_1(r,t)\cdot dS = \iint_S f_2(r,t)\cdot dS$$

The left hand side is the flow rate of quantity q occurring inside the closed surface S. This must be equal to

$$\iint_S f_1(r,t)\cdot dS = \Sigma(t) - \frac{dq(t)}{dt}$$

since some is produced by sources, hence the positive term Σ, but some is also leaking out by passing through the surface, implied by the negative term $-dq/dt$. Similarly the right hand side is the amount of flux passing through the surface and out of it, so

$$\iint_S f_2(r,t)\cdot dS = \oiint sf(r,t)\cdot dS$$

Equating these:

$$\Sigma(t) - \frac{dq(t)}{dt} = \oiint sf(r,t)\cdot dS$$

$$\frac{dq}{dt} + \oiint sf\cdot dS = \Sigma$$

which is the integral form again.

Electromagnetism

3-currents: In electromagnetic theory, the continuity equation can either be regarded as an empirical law expressing (local) charge conservation, or can be *derived* as a consequence of two of Maxwell's equations. It states that the divergence of the current density J (in amperes per square metre) is equal to the negative rate of change of the charge density ρ (in coulombs per cubic metre),

$$\nabla \cdot J = -\frac{\partial \rho}{\partial t}$$

Maxwell's equations are a quick way to obtain the continuity of charge.

Consistency with Maxwell's Equations

4-currents

Conservation of a current (not necessarily an electromagnetic current) is expressed compactly as the Lorentz invariant divergence of a four-current:

$$J^{\mu} = (c\rho, j)$$

where

- c is the speed of light
- ρ; the charge density
- j the conventional 3-current density.
- μ; labels the space-time dimension

since $$\partial_{\mu} J^{\mu} = \frac{\partial \rho}{\partial t} + \nabla \cdot j$$

then $$\partial_{\mu} J^{\mu} = 0$$

which implies that the current is conserved:

$$\frac{\partial \rho}{\partial t} + \nabla \cdot j = 0.$$

Interpretation

Current is the movement of charge. The continuity equation says that if charge is moving out of a differential volume (i.e. divergence of current density is positive) then the amount of charge within that volume is going to decrease, so the rate of change of charge density is negative. Therefore the continuity equation amounts to a conservation of charge.

Fluid Dynamics

In fluid dynamics, the continuity equation states that, in any steady state process, the rate at which mass enters a system is equal to the rate at which mass leaves the system. In fluid dynamics, the continuity equation is analogous to Kirchhoff's current law in electric circuits.

The differential form of the continuity equation is:

$$\frac{\partial \rho}{\partial t} + \nabla \cdot (\rho u) = 0$$

where

- ρ is fluid density,
- t is time,
- u is the flow velocity vector field.

If ρ is a constant, as in the case of incompressible flow, the mass continuity equation simplifies to a volume continuity equation:

$$\nabla \cdot u = 0,$$

which means that the divergence of velocity field is zero everywhere. Physically, this is equivalent to saying that the local volume dilation rate is zero.

Further, the Navier-Stokes equations form a vector continuity equation describing the conservation of linear momentum.

Energy

Conservation of energy (which, in non-relativistic situations, can only be transferred, and not created or destroyed) leads to a continuity equation, an alternative mathematical statement of energy conservation to the thermodynamic laws.

Letting:

- u = local energy density (energy per unit volume),
- q = energy flux (transfer of energy per unit cross-sectional area per unit time) as a vector,

the continuity equation is:

$$\nabla \cdot q + \frac{\partial u}{\partial t} = 0$$

Quantum Mechanics

In quantum mechanics, the *conservation of probability* also yields a continuity equation. The terms in the equation require these definitions, and are slightly less obvious than the other forms of volume densities, currents, current densities etc., so they are outlined here:

- The wavefunction Ψ for a single particle in the position-time space (rather than momentum space) - i.e. functions of position r and time t, $\Psi = \Psi(\mathrm{r}, t) = \Psi(x, y, z, t)$.
- The probability density function $\rho = \rho(\mathrm{r}, t)$ is:

$$\rho = \Psi^* \Psi$$

- The probability that a measurement of the particle's position will yield a value within V at t, denoted by $P = P_{\mathrm{r}=V}(t)$, is:

$$P = P_{r \in V}(t) = \int_V \Psi^* \Psi \mathrm{d}V = \int_V |\Psi|^2 \,\mathrm{d}V$$

- The probability current (aka probability flux) j:

$$j = \frac{\hbar}{2mi}\left[\Psi^*\left(\nabla\Psi\right) - \Psi\left(\nabla\Psi^*\right)\right].$$

With these definitions the continuity equation reads:

$$\nabla \cdot j + \frac{\partial \rho}{\partial t} = 0 \rightleftharpoons \nabla \cdot j + \frac{\partial |\Psi|^2}{\partial t} = 0.$$

Either form is usually quoted. Intuitively; the above quantities indicate this represents the flow of probability. The chance of finding the particle at some r t flows like a fluid, the particle itself does not flow deterministically in the same vector field.

Consistency with Schrödinger's Equation

The 3-d time dependent Schrödinger equation and its complex conjugate ($i \to -i$) throughout are respectively:

$$-\frac{\hbar^2}{2m}\nabla^2\Psi + U\Psi = i\hbar\frac{\partial \Psi}{\partial t},$$

$$-\frac{\hbar^2}{2m}\nabla^2\Psi^* + U\Psi^* = -i\hbar\frac{\partial \Psi^*}{\partial t},$$

where U is the potential function. The partial derivative of ρ with respect to t is:

$$\frac{\partial \rho}{\partial t} = \frac{\partial |\Psi|^2}{\partial t} = \frac{\partial}{\partial t}\left(\Psi^*\Psi\right) - \Psi^*\frac{\partial \Psi}{\partial t} + \Psi\frac{\partial \Psi^*}{\partial t}.$$

Multiplying the Schrödinger equation by Ψ^* then solving for $\Psi^* \partial\Psi / \partial t$, and similarly multiplying the complex conjugated Schrödinger equation by Ψ then solving for $\Psi \partial\Psi^* / \partial t$;

$$\Psi^* \frac{\partial \Psi}{\partial t} = \frac{1}{i\hbar}\left[-\frac{\hbar^2 \Psi^*}{2m}\nabla^2\Psi + U\Psi^*\Psi\right],$$

$$\Psi \frac{\partial \Psi^*}{\partial t} = -\frac{1}{i\hbar}\left[-\frac{\hbar^2 \Psi}{2m}\nabla^2\Psi^* + U\Psi\Psi^*\right],$$

substituting into the time derivative of ρ:

$$\frac{\partial \rho}{\partial t} = \frac{1}{i\hbar}\left[-\frac{\hbar^2 \Psi^*}{2m}\nabla^2\Psi + U\Psi^*\Psi\right] - \frac{1}{i\hbar}\left[-\frac{\hbar^2 \Psi}{2m}\nabla^2\Psi^* + U\Psi\Psi^*\right]$$

$$= \frac{\hbar}{2im}\left[\Psi\nabla^2\Psi^* - \Psi^*\nabla^2\Psi\right]$$

The Laplacian operators (∇^2) in the above result suggest that the right hand side is the divergence of j, and the reversed order of terms imply this is the negative of j, altogether:

$$\nabla\cdot j = \nabla\cdot\left[\frac{\hbar}{2mi}\left(\Psi^*\left(\nabla\Psi\right) - \Psi\left(\nabla\Psi^*\right)\right)\right]$$

$$= \frac{\hbar}{2mi}\left[\Psi^*\left(\nabla^2\Psi\right) - \Psi\left(\nabla^2\Psi^*\right)\right]$$

$$= -\frac{\hbar}{2mi}\left[\Psi\left(\nabla^2\Psi^*\right) - \Psi^*\left(\nabla^2\Psi\right)\right]$$

so the continuity equation is:

$$\frac{\partial \rho}{\partial t} = -\nabla\cdot j$$

$$\frac{\partial \rho}{\partial t} + \nabla\cdot j = 0$$

The integral form follows as for the general equation.

Consistency with the Wavefunction Probability Distribution

The time derivative of P is

$$\frac{dP}{dt} = \frac{\partial}{\partial t}\iiint_V |\Psi|^2\, dV = \iiint_V \frac{\partial |\Psi|^2}{\partial t} dV = \iiint_V \left(\frac{\partial \Psi}{\partial t}\Psi^* + \Psi\frac{\partial \Psi^*}{\partial t}\right) dV$$

where the last equality follows from the product rule and the factthat the shape of V is fixed for the calculation and therefore independent of time - i.e. the time derivative can be moved through the integral.

To simplify this further consider again the time dependent Schrödinger equation and its complex conjugate, in terms of the time derivatives of Ψ and Ψ^* respectively:

$$\frac{\partial \Psi}{\partial t} = \frac{i\hbar}{2m}\nabla^2\Psi - \frac{i}{\hbar}U\Psi$$

$$\frac{\partial \Psi^*}{\partial t} = -\left[\frac{i\hbar}{2m}\nabla^2\Psi^* - \frac{i}{\hbar}U\Psi^*\right]$$

Substituting into the preceding equation:

$$\int_V \frac{\partial |\Psi|^2}{\partial t}\mathrm{d}V = -\iiint_V \frac{\hbar}{2mi}\left(\Psi^*\nabla^2\Psi - \Psi\nabla^2\Psi^*\right)\mathrm{d}V.$$

From the product rule for the divergence operator

$$\begin{aligned}\nabla\cdot\left(\Psi^*\nabla\Psi - \Psi\nabla\Psi^*\right) &= \nabla\Psi^*\cdot\nabla\Psi + \Psi^*\nabla^2\Psi - \nabla\Psi\cdot\nabla\Psi^* - \Psi\nabla^2\Psi^* \\ &= \Psi^*\nabla^2\Psi - \Psi\nabla^2\Psi^*\end{aligned}$$

Substituting:

$$\iiint_V \frac{\partial |\Psi|^2}{\partial t}\mathrm{d}V = -\iiint_V \nabla\cdot\left[\frac{\hbar}{2mi}\left(\Psi^*\nabla\Psi - \Psi\nabla\Psi^*\right)\right]\mathrm{d}V$$

On the right side, the argument of the divergence operator is j,

$$\iiint_V \frac{\partial |\Psi|^2}{\partial t}\mathrm{d}V = -\iiint_V (\nabla\cdot j)\mathrm{d}V$$

using the divergence theorem again gives the integral form:

$$\iiint_V \frac{\partial |\Psi|^2}{\partial t}\mathrm{d}V = -\oiint S(j\cdot\hat{n})\mathrm{d}S$$

To obtain the differential form:

$$\iiint_V \left(\frac{\partial |\Psi|^2}{\partial t} + \nabla\cdot j\right)\mathrm{d}V = 0$$

The differential form follows from the fact that the preceding equation holds for *all* V, and as the integrand is a continuous function of space, it must vanish everywhere:

$$\frac{\partial |\Psi|^2}{\partial t} + \nabla \cdot j = 0.$$

Inverse-square Laws

Any *inverse-square law* can instead be written in a *Gauss' law*-type form (with a differential and integral form, as described above).

Two examples are Gauss' law (in electrostatics), which follows from the inverse-square Coulomb's law, and Gauss' law for gravity, which follows from the inverse-square Newton's law of universal gravitation. The derivation of the Gauss' law-type equation from the inverse-square formulation (or vice-versa) is exactly the same in both cases.

History

The theorem was first discovered by Lagrange in 1762, then later independently rediscovered by Gauss in 1813, by Green in 1825 and in 1831 by Ostrogradsky, who also gave the first proof of the theorem. Subsequently, variations on the divergence theorem are correctly called Ostrogradsky's theorem, but also commonly Gauss's theorem, or Green's theorem.

Examples

To verify the planar variant of the divergence theorem for a region *R*, where

$$F(x,y) = 2yi + 5xj,$$

and *R* is the region bounded by the circle

$$x^2 + y^2 = 1.$$

The boundary of *R* is the unit circle, *C*, that can be represented parametrically by:

$$x = \cos(s);\ y = \sin(s)$$

such that $0 \le s \le 2\pi$ where *s* units is the length arc from the point $s = 0$ to the point *P* on *C*. Then a vector equation of *C* is

$$C(s) = \cos(s)i + \sin(s)j.$$

At a point P on *C*:

$$P = (\cos(s), \sin(s)) \Rightarrow F = 2\sin(s)i + 5\cos(s)j.$$

Therefore,

$$\begin{aligned}\oint_C F\cdot N ds &= \int_0^{2\pi}(2\sin s i + 5\cos s j)\cdot(\cos s i + \sin s j)ds \\ &= \int_0^{2\pi}(2\sin s\cos s + 5\sin s\cos s)ds \\ &= 7\int_0^{2\pi}\sin s\cos s ds \\ &= 0.\end{aligned}$$

Because $M=2y, \frac{\partial M}{\partial x}=0$, and because $N=5x, \frac{\partial N}{\partial y}=0$. Thus

$$\iint_R \operatorname{div} F dA = \iint_R \left(\frac{\partial M}{\partial x}+\frac{\partial N}{\partial y}\right)dA = 0.$$

Generalization to Tensor Fields

Writing the theorem in index notation:

$$\iiint_V \frac{\partial F_i}{\partial x_i}dV = \oiint S F_i n_i dS$$

suggestively, replacing the vector field F with a rank-n tensor field T, this can be generalized to:

$$\iiint_V \frac{\partial T_{i_1 i_2 \cdots i_q \cdots i_n}}{\partial x_{i_q}}dV = \oiint S T_{i_1 i_2 \cdots i_q \cdots i_n} n_{i_q} dS.$$

where on each side, tensor contraction occurs for at least one index. This form of the theorem is still in 3d, each index takes values 1, 2, and 3. It can be generalized further still to higher dimensions (for example to 4d spacetime in general relativity).

Advection

In chemistry, engineering and earth sciences, advection is a transport mechanism of a substance or conserved property by a fluid due to the fluid's bulk motion. An example of advection is the transport of pollutants or silt in a river by bulk water flow downstream. Another commonly advected quantity is energy or enthalpy. Here the fluid may be any material that contains thermal energy, such as water or air. In general, any substance or conserved, extensive quantity can be advected by a fluid that can hold or contain the quantity or substance.

In advection, a fluid transports some conserved quantity or material via bulk motion. The fluid's motion is described mathematically as a vector field, and the transported material is described by a scalar field showing its distribution over space. Advection requires currents in the fluid, and so cannot happen in rigid solids.

It does not include transport of substances by simple diffusion. Advection is sometimes confused with the more encompassing process of convection. In fact, convective transport is the sum of advective transport and diffusive transport.

In meteorology and physical oceanography, advection often refers to the transport of some property of the atmosphere or ocean, such as heat, humidity or salinity. Advection is important for the formation of orographic clouds and the precipitation of water from clouds, as part of the hydrological cycle.

Distinction Between Advection and Convection

The term *advection* is sometimes used as a synonym for *convection*. Technically, convection is the sum of transport by diffusion and advection. Advective transport describes the movement of some quantity via the bulk flow of a fluid (as in a river or pipeline).

Meteorology

In meteorology and physical oceanography, advection often refers to the transport of some property of the atmosphere or ocean, such as heat, humidity or salinity. Advection is important for the formation of orographic clouds and the precipitation of water from clouds, as part of the hydrological cycle.

Other Quantities

The advection equation also applies if the quantity being advected is represented by a probability density function at each point, although accounting for diffusion is more difficult.

Mathematics of Advection

The advection equation is the partial differential equation that governs the motion of a conserved scalar field as it is advected by a known velocity vector field. It is derived using the scalar field's conservation law, together with Gauss's theorem, and taking the infinitesimal limit.

One easily-visualized example of advection is the transport of ink dumped into a river. As the river flows, ink will move downstream in a "pulse" via advection, as the water's movement itself transports the ink.

If added to a lake without significant bulk water flow, the ink would simply disperse outwards from its source in a diffusive manner, which is not advection. Note that as it moves downstream, the "pulse" of ink will also spread via diffusion. The sum of these processes is called convection.

The Advection Equation

In Cartesian coordinates the advection operator is

$$u\cdot\nabla = u_x \frac{\partial}{\partial x} + u_y \frac{\partial}{\partial y} + u_z \frac{\partial}{\partial z}.$$

where u = (u_x, u_y, u_z) is the velocity field, and ∇ is the del operator (note that Cartesian coordinates are used here).

The advection equation for a conserved quantity described by a scalar field ψ is expressed mathematically by a continuity equation:

$$\frac{\partial\psi}{\partial t} + \nabla\cdot(\psi u) = 0$$

where ∇ is the divergence operator and again u is the velocity vector field. Frequently, it is assumed that the flow is incompressible, that is, the velocity field satisfies

$$\nabla\cdot u = 0$$

and u is said to be solenoidal. If this is so, the above equation reduces to

$$\frac{\partial\psi}{\partial t} + u\cdot\nabla\psi = 0.$$

In particular, if the flow is steady, then

$$u\cdot\nabla\psi = 0$$

which shows that ψ is constant along a streamline.

If a vector quantity a (such as a magnetic field) is being advected by the solenoidal velocity field u, the advection equation above becomes:

$$\frac{\partial a}{\partial t}+(u\cdot\nabla)a=0.$$

Here, a is a vector field instead of the scalar field ψ}.

Solving the Equation

The advection equation is not simple to solve numerically: the system is a hyperbolic partial differential equation, and interest typically centres on discontinuous "shock" solutions (which are notoriously difficult for numerical schemes to handle).

Even with one space dimension and a constant velocity field, the system remains difficult to simulate. The equation becomes

$$\frac{\partial\psi}{\partial t}+u_x\frac{\partial\psi}{\partial x}=0$$

where $\psi = \psi(x, t)$ is the scalar field being advected and u_x is the x component of the vector u = $(u_x,0,0)$.

According to Zang, numerical simulation can be aided by considering the skew symmetric form for the advection operator.

$$\frac{1}{2}u\cdot\nabla u+\frac{1}{2}\nabla(uu)$$

where $\nabla(uu)=[\nabla(uu_x),\nabla(uu_y),\nabla(uu_z)]$

and u is the same as above.

Since skew symmetry implies only imaginary eigenvalues, this form reduces the "blow up" and "spectral blocking" often experienced in numerical solutions with sharp discontinuities.

Using vector calculus identities, these operators can also be expressed in other ways, available in more software packages for more coordinate systems.

$$u\cdot\nabla u=\nabla\left(\frac{||u||^2}{2}\right)+(\nabla\times u)\times u$$

$$\frac{1}{2}u\cdot\nabla u+\frac{1}{2}\nabla(uu)=\nabla\left(\frac{||u||^2}{2}\right)+(\nabla\times u)\times u+\frac{1}{2}u(\nabla\cdot u)$$

This form also makes visible that the skew symmetric operator introduces error when the velocity field diverges.

Beltrami Vector Field

In vector calculus, a Beltrami vector field, named after Eugenio Beltrami, is a vector field in three dimensions that is parallel to its own curl. That is, F is a Beltrami vector field provided that

$$F \times (\nabla \times F) = 0.$$

If F is solenoidal - that is, if $\nabla . F = 0$ such as for an incompressible fluid or a magnetic field, we may examine $\nabla \times (\nabla \times F) \equiv \nabla^2 F - \nabla \nabla . F$ and use apply this identity twice to find that

$$\nabla^2 F = \nabla \times (\lambda F)$$

and if we further assume that $\lambda$is a constant, we arrive at the simple form

$$\nabla^2 F = \lambda^2 F.$$

Beltrami vector fields with nonzero curl correspond to Euclidean contact forms in three dimensions.

The vector field

$$F = -\frac{z}{\sqrt{1+z^2}} i + \frac{1}{\sqrt{1+z^2}} j$$

is a multiple of the standard contact structure -*z*i + j, and furnishes an example of a Beltrami vector field.

Bivector

In mathematics, a bivector or 2-vector is a quantity in geometric algebra or exterior algebra that generalises the idea of a vector. If a scalar is considered an order zero quantity, and a vector is an order one quantity, then a bivector can be thought of as being of order two. Bivectors have applications in many areas of mathematics and physics. They are related to complex numbers in two dimensions and to both pseudovectors and quaternions in three dimensions. They can be used to generate rotations in any dimension, and are a useful tool for classifying such rotations. They also are used in physics, tying together a number of otherwise unrelated quantities.

Bivectors are generated by the exterior product on vectors – given two vectors a and b their exterior product a ≠ b is a bivector. But not all bivectors can be generated this way, and in higher dimensions a sum of exterior products is often needed. More precisely a bivector that requires only a single exterior product is *simple*; in two and three dimensions all bivectors are simple, but in higher dimensions this is not generally the case. The exterior product is antisymmetric, so b ≠ a is the negation of the bivector a ≠ b, producing a rotation with the opposite sense, and a ≠ a is the zero bivector.

Geometrically, a simple bivector can be interpreted as an oriented plane segment, much as vectors can be thought of as directed line segments. Specifically for the bivector a ≠ b, its *magnitude* is the area of the parallelogram with edges a and b, its *attitude* that of any plane specified by a and b, and its *orientation* the sense of the rotation that would align a with b. It does not have a definite location or position.

Formal Definition

For this article the bivector will be considered only in real geometric algebras. This in practice is not much of a restriction, as all useful applications are drawn from such algebras. Also unless otherwise stated, all examples have a Euclidean metric and so a quadratic form with signature 1.

Geometric Algebra and the Geometric Product

The bivector arises from the definition of the geometric product over a vector space. For vectors a, b and c, the geometric product on vectors is defined as follows:

Associativity:

$$(ab)c = a(bc)$$

Left and right distributivity:

$$a(b+c) = ab + ac$$

$$(b+c)a = ba + ca$$

Contraction:

$$a^2 = Q(a) = \epsilon_a |a|^2$$

Where *Q* is the quadratic form, | a | is the magnitude of a and υ_a is the signature of the vector. For a space with Euclidean metric υ_a is 1 so can be omitted, and the contraction condition becomes:

$$a^2 = |a|^2$$

The Interior Product

From associativity a(ab) = a^2b, a scalar times b. When b is not parallel to and hence not a scalar multiple of a, ab cannot be a scalar. But

$$\frac{1}{2}(ab+ba) = \frac{1}{2}((a+b)^2 - a^2 - b^2)$$

is a sum of scalars and so a scalar. From the law of cosines on the triangle formed by the vectors its value is |a| |b|cosθ, where θ is the angle between the vectors. It is therefore identical to the interior product between two vectors, and is written the same way,

$$a{\cdot}b = \frac{1}{2}(ab+ba).$$

It is symmetric, scalar valued, and can be used to determine the angle between two vectors: in particular if a and b are orthogonal the product is zero.

The Exterior Product

In the same way another quantity can be written down:

$$\frac{1}{2}(ab-ba)$$

This is called the exterior product, a ≠ b. It is antisymmetric in a and b, that is

$$a \wedge b = \frac{1}{2}(ab-ba) = -b \wedge a$$

By addition:

$$a{\cdot}b + a \wedge b = \frac{1}{2}(ab+ba) + \frac{1}{2}(ab-ba) = ab$$

That is the geometric product is the sum of the symmetric interior product and antisymmetric exterior product.

To calculate a ≠ b consider the sum

$$(a{\cdot}b)^2 - (a \wedge b)^2$$

Expanding using the geometric product and simplifying gives

$$(a \cdot b)^2 - (a \wedge b)^2 = a^2 b^2$$

So using the Pythagorean trigonometric identity:

$$(a \wedge b)^2 = (a \cdot b)^2 - a^2 b^2 = |a|^2 |b|^2 (\cos^2 \theta - 1) = -|a|^2 |b|^2 \sin^2 \theta$$

With a negative square it cannot be a scalar or vector quantity, so it is a new sort of object, a bivector. It has magnitude $|a||b|\sin\theta$, where θ is the angle between the vectors, and so is zero for parallel vectors.

To distinguish them from vectors, bivectors are written here with bold capitals, for example:

$$A = a \wedge b = -b \wedge a,$$

although other conventions are used, in particular as vectors and bivectors are both elements of the geometric algebra.

Properties

The space $\wedge^2 \mathbb{R}^{n}$: the algebra generated by the geometric product is the geometric algebra over the vector space. For a Euclidean vector space it is written $\mathcal{G}_n$ or $C\ell_n(\mathbb{R})$, where n is the dimension of the vector space $\mathbb{R}^n$. $C\ell_n$ is both a vector space and an algebra, generated by all the products between vectors in $\mathbb{R}^n$, so it contains all vectors and bivectors.

More precisely as a vector space it contains the vectors and bivectors as subspaces. The space of all bivectors is written $\wedge^2 \mathbb{R}^n$.Unlike $\mathbb{R}^n$ it is not a Euclidean subspace; nor is it a subalgebra.

The Even Subalgebra

The subalgebra generated by the bivectors is the *even subalgebra* of the geometric algebra, written $C\ell$ +

n . This algebra results from considering all products of scalars and bivectors generated by the geometric product. It has dimension 2^{n-1}, and contains $\wedge^2 \mathbb{R}^n$ as a linear subspace with dimension ¹D ${}_2n(n - 1)$ (a triangular number). In two and three dimensions the even subalgebra contains only scalars and bivectors, and each is of particular interest. In two dimensions the even subalgebra is isomorphic to the complex numbers, ℓ, while in three it is isomorphic to the quaternions,

ℓ. More generally the even subalgebra can be used to generate rotations in any dimension, and can be generated by bivectors in the algebra.

Magnitude

As noted in the previous section the magnitude of a simple bivector, that is one that is the exterior product of two vectors a and b, is $|a||b|\sin\theta$, where θ is the angle between the vectors. It is written $|B|$, where B is the bivector.

For general bivectors the magnitude can be calculated by taking the norm of the bivector considered as a vector in the space $\wedge^2 \mathbb{R}^n$. If the magnitude is zero then all the bivector's components are zero, and the bivector is the zero bivector which as an element of the geometric algebra equals the scalar zero.

Unit Bivectors

A unit bivector is one with unit magnitude. It can be derived from any non-zero bivector by dividing the bivector by its magnitude, that is

$$\frac{B}{|B|}.$$

Of particular interest are the unit bivectors formed from the products of the standard basis. If e_i and e_j are distinct basis vectors then the product $e_i \neq e_j$ is a bivector. As the vectors are orthogonal this is just e_ie_j, written e_{ij}, with unit magnitude as the vectors are unit vectors. The set of all such bivectors form a basis for $\wedge_2 \mathbb{R}_n$. For instance in four dimensions the basis for $\wedge_2 \mathbb{R}_4$ is (e_1e_2, e_1e_3, e_1e_4, e_2e_3, e_2e_4, e_3e_4) or (e_{12}, e_{13}, e_{14}, e_{23}, e_{24}, e_{34}).

Simple Bivectors

The exterior product of two vectors is a bivector, but not all bivectors are exterior products of two vectors. For example in four dimensions the bivector

$$B = e_1 \wedge e_2 + e_3 \wedge e_4 = e_1e_2 + e_3e_4 = e_{12} + e_{34}$$

cannot be written as the exterior product of two vectors. A bivector that can be written as the exterior product of two vectors is simple. In two and three dimensions all bivectors are simple, but not in four or more dimensions; in four dimensions every bivector is the sum of at most two exterior products. A bivector has a real square if and only if it is simple, and only simple bivectors can be represented geometrically by a oriented plane area.

Product of two Bivectors

The geometric product of two bivectors, A and B, is

$$AB = A \cdot B + A \times B + A \wedge B.$$

The quantity A . B is the scalar valued interior product, while A ≠ B is the grade 4 exterior product that arises in four or more dimensions. The quantity A × B is the bivector valued commutator product, given by

$$A \times B = \frac{1}{2}(AB - BA),$$

The space of bivectors $\wedge^2 \mathbb{R}^n$ are a Lie algebra over $\mathbb{R}$, with the commutator product as the Lie bracket. The full geometric product of bivectors generates the even subalgebra.

Of particular interest is the product of a bivector with itself. As the commutator product is antisymmetric the product simplifies to

$$AA = A \cdot A + A \wedge A.$$

If the bivector is *simple* the last term is zero and the product is the scalar valued A · A, which can be used as a check for simplicity.

In particular the exterior product of bivectors only exists in four or more dimensions, so all bivectors in two and three dimensions are simple.

Two Dimensions

When working with coordinates in geometric algebra it is usual to write the basis vectors as (e_1, e_2, ...), a convention that will be used here.

A vector in real two dimensional space $\mathbb{R}^2$ can be written $a = a_1e_1 + a_2e_2$, where a_1 and a_2 are real numbers, e_1 and e_2 are orthonormal basis vectors. The geometric product of two such vectors is

$$\begin{aligned} ab &= (a_1e_1 + a_2e_2)(b_1e_1 + b_2e_2) \\ &= a_1b_1e_1e_1 + a_1b_2e_1e_2 + a_2b_1e_2e_1 + a_2b_2e_2e_2 \\ &= a_1b_1 + a_2b_2 + (a_1b_2 - a_2b_1)e_1e_2. \end{aligned}$$

This can be split into the symmetric, scalar valued, interior product and an antisymmetric, bivector valued exterior product:

$$\begin{aligned} a{\cdot}b &= a_1b_1 + a_2b_2, \\ a \wedge b &= (a_1b_2 - a_2b_1)e_1e_2 = (a_1b_2 - a_2b_1)e_{12}. \end{aligned}$$

All bivectors in two dimensions are of this form, that is multiples of the bivector e_1e_2, written e_{12} to emphasise it is a bivector rather than a vector. The magnitude of e_{12} is 1, with

$$e_{12}^2 = -1,$$

so it is called the unit bivector. The term unit bivector can be used in other dimensions but it is only uniquely defined in two dimensions and all bivectors are multiples of e_{12}. As the highest grade element of the algebra e_{12} is also the pseudoscalar which is given the symbol i.

Complex Numbers

With the properties of negative square and unit magnitude the unit bivector can be identified with the imaginary unit from complex numbers. The bivectors and scalars together form the even subalgebra of the geometric algebra, which is isomorphic to the complex numbers $\mathbb{C}$. The even subalgebra has basis $(1, e_{12})$, the whole algebra has basis $(1, e_1, e_2, e_{12})$.

The complex numbers are usually identified with the coordinate axes and two dimensional vectors, which would mean associating them with the vector elements of the geometric algebra. There is no contradiction in this, as to get from a general vector to a complex number an axis needs to be identified as the real axis, e_1 say. This multiplies by all vectors to generate the elements of even subalgebra. All the properties of complex numbers can be derived from bivectors, but two are of particular interest. First as with complex numbers products of bivectors and so the even subalgebra are commutative. This is only true in two dimensions, so properties of the bivector in two dimensions that depend on commutativity do not usually generalise to higher dimensions.

Second a general bivector can be written

$$\theta e_{12} = i\theta,$$

where θ is a real number. Putting this into the Taylor series for the exponential map and using the property $e_{12}{}^2 = -1$ results in a bivector version of Euler's formula,

$$e^{\theta e_{12}} = e^{i\theta} = \cos\theta + i\sin\theta,$$

which when multiplied by any vector rotates it through an angle θ about the origin:

$$(x'e_1 + y'e_2) = (xe_1 + ye_2)e^{i\theta}.$$

The product of a vector with a bivector in two dimensions is anticommutative, so the following products all generate the same rotation

$$v' = ve^{i\theta} = e^{-i\theta}v = e^{\frac{-i\theta}{2}} ve^{\frac{i\theta}{2}}.$$

Of these the last product is the one that generalises into higher dimensions. The quantity needed is called a rotor and is given the symbol R, so in two dimensions a rotor that rotates through angle θ can be written

$$R = e^{\frac{-i\theta}{2}} = e^{\frac{-\theta e_{12}}{2}},$$

and the rotation it generates is

$$v' = RvR^{-1}.$$

Quaternions

Bivectors are not closed under the geometric product, but the even subalgebra is. In three dimensions it consists of all scalar and bivector elements of the geometric algebra, so a general element can be written for example a + A, where a is the scalar part and A is the bivector part. It is written $C\ell_3^+$ and has basis (1, e_{23}, e_{31}, e_{12}). The product of two general elements of the even subalgebra is

$$(a + A)(b + B) = ab + aB + bA + A\cdot B + A \times B.$$

The even subalgebra, that is the algebra consisting of scalars and bivectors, is isomorphic to the quaternions, $\mathbb{H}$. This can be seen by comparing the basis to the quaternion basis, or from the above product which is identical to the quaternion product, except for a change of sign which relates to the negative products in the bivector interior product A · B. Other quaternion properties can be similarly related to or derived from geometric algebra.

This suggests that the usual split of a quaternion into scalar and vector parts would be better represented as a split into scalar and bivector parts; if this is done there is no special quaternion product, there is just the normal geometric product on the elements.

It also relates quaternions in three dimensions to complex numbers in two, as each is isomorphic to the even subalgebra for the dimension, a relationship that generalises to higher dimensions.

Rotation Vector

The rotation vector, from the axis angle representation of rotations, is a compact way of representing rotations in three dimensions. In its most compact form it consists of a vector, the product of the a unit vector that is the axis of rotation and the angle of rotation, so the magnitude of the vector is the rotation angle.

In geometric algebra this vector is classified as a bivector. This can be seen in its relation to quaternions. If the axis is ω and the angle of rotation is θ then the rotation vector is $\omega\theta$ quaternion associated with the rotation is

$$q = \left(\cos\left(\frac{\theta}{2}\right), \omega\sin\left(\frac{\theta}{2}\right)\right)$$

but this is just the exponent of half of the bivector $\Omega\theta$, that is

$$e^{\frac{\Omega\theta}{2}} = \cos\left(\frac{\theta}{2}\right) + \Omega\sin\left(\frac{\theta}{2}\right)$$

So rotation vectors are bivectors, just as quaternions are elements of the geometric algebra, and they are related by the exponential map in that algebra.

Rotors

The bivector $\Omega\theta$ generates a rotation through the exponential map. The even elements generated rotate a general vector in three dimensions in the same way as quaternions:

$$v' = e^{\frac{\Omega\theta}{2}} v e^{-\frac{\Omega\theta}{2}}.$$

As to two dimensions the quantity $e^{\Omega\theta}$ is called a rotor and written R. The quantity $e^{-\Omega\theta}$ is then R^{-1}, and they generate rotations as follows

$$v' = RvR^{-1}.$$

This is identical to two dimensions, except here rotors are four-dimensional objects isomorphic to the quaternions. This can be generalised to all dimensions, with rotors, elements of the even subalgebra with unit magnitude, being generated by the exponential map from bivectors. They form a double cover over the rotation group, so the rotors R and $-R$ represent the same rotation.

Matrices

Bivectors are isomorphic to skew-symmetric matrices; the general bivector $B_{23}e_{23} + B_{31}e_{31} + B_{12}e_{12}$ maps to the matrix

$$M_B = \begin{pmatrix} 0 & -B_{12} & B_{31} \\ B_{12} & 0 & -B_{23} \\ -B_{31} & B_{23} & 0 \end{pmatrix}.$$

This multiplied by vectors on both sides gives the same vector as the product of a vector and bivector; an example is the angular velocity tensor.

Skew symmetric matrices generate orthogonal matrices with determinant 1 through the exponential map. In particular the exponent of a bivector associated with a rotation is a rotation matrix, that is the rotation matrix M_R given by the above skew-symmetric matrix is

$$M_R = e^{M_B}.$$

The rotation described by M_R is the same as that described by the rotor R given by

$$R = e^{\frac{B}{2}},$$

and the matrix M_R can be also calculated directly from rotor R:

$$M_R = \begin{pmatrix} (Re_1R^{-1})\cdot e_1 & (Re_2R^{-1})\cdot e_1 & (Re_3R^{-1})\cdot e_1 \\ (Re_1R^{-1})\cdot e_2 & (Re_2R^{-1})\cdot e_2 & (Re_3R^{-1})\cdot e_2 \\ (Re_1R^{-1})\cdot e_3 & (Re_2R^{-1})\cdot e_3 & (Re_3R^{-1})\cdot e_3 \end{pmatrix}.$$

Bivectors are related to the eigenvalues of a rotation matrix. Given a rotation matrix M the eigenvalues can calculated by solving the characteristic equation for that matrix $0 = \det(M - \lambda I)$. By the fundamental theorem of algebra this has three roots, but only one real root as there is only one eigenvector, the axis of rotation. The other roots must be a complex conjugate pair.

They have unit magnitude so purely imaginary logarithms, equal to the magnitude of the bivector associated with the rotation, which is also the angle of rotation. The eigenvectors associated with the complex eigenvalues are in the plane of the bivector, so the exterior product of two non-parallel eigenvectors result in the bivector, or at least a multiple of it.

Axial Vectors

The rotation vector is an example of an axial vector. Axial vectors or pseudovectors are vectors that undergo a sign change compared to normal or polar vectors under inversion, that is when reflected or otherwise inverted. Examples include quantities like torque, angular momentum and vector magnetic fields. Such quantities can be described as bivectors in geometric algebra; that is quantities that might use axial vectors in vector algebra are better represented by bivectors in geometric algebra.

More precisely, the Hodge dual gives the isomorphism between axial vectors and bivectors, so each axial vector is associated with a bivector and vice-versa; that is

$$A = {*a}; \quad a = {*A}$$

where * indicates the Hodge dual. Alternately, using the unit pseudoscalar in $C\ell_3(\mathbb{R})$, $i = e_1e_2e_3$ gives

$$A = ai; \quad a = -Ai.$$

This is easier to use as the product is just the geometric product. But it is antisymmetric because (as in two dimensions) the unit pseudoscalar i squares to -1, so a negative is needed in one of the products.

This relationship extends to operations like the vector valued cross product and bivector valued exterior product, as when written as determinants they are calculated in the same way:

$$a \times b = \begin{vmatrix} e_1 & e_2 & e_3 \\ a_1 & a_2 & a_3 \\ b_1 & b_2 & b_3 \end{vmatrix}; \quad a \wedge b = \begin{vmatrix} e_{23} & e_{31} & e_{12} \\ a_1 & a_2 & a_3 \\ b_1 & b_2 & b_3 \end{vmatrix},$$

so are related by the Hodge dual:

$$*(a \wedge b) = a \times b \ ; \ *(a \times b) = a \wedge b.$$

Bivectors have a number of advantages over axial vectors. They better disambiguate axial and polar vectors, that is the quantities represented by them, so it is clearer which operations are allowed and what their results are.

For example the inner product of a polar vector and an axial vector resulting from the cross product in the triple product should result in a pseudoscalar, a result which is more obvious if the calculation is framed as the exterior product of a vector and bivector.

They generalises to other dimensions; in particular bivectors can be used to describe quantities like torque and angular momentum in two as well as three dimensions. Also, they closely match geometric intuition in a number of ways.

Geometric Interpretation

As suggested by their name and that of the algebra, one of the attractions of bivectors is that they have a natural geometric interpretation.

This can be described in any dimension but is best done in three where parallels can be drawn with more familiar objects, before being applied to higher dimensions. In two dimensions the geometric interpretation is trivial, as the space is two dimensional so has only one plane, and all bivectors are associated with it differing only by a scale factor.

All bivectors can be interpreted as planes, or more precisely as directed plane segments. In three dimensions there are three properties of a bivector that can be interpreted geometrically:

- The arrangement of the plane in space, precisely the attitude of the plane (or alternately the rotation, geometric orientation or gradient of the plane), is associated with the ratio of the bivector components. In particular the three basis bivectors, e_{23}, e_{31} and e_{12}, or scalar multiples of them, are associated with the *yz*-plane, *xz*-plane and *xy*-plane respectively.
- The magnitude of the bivector is associated with the area of the plane segment. The area does not have a particular shape so any shape can be used. It can even be represented in other ways, such as by an angular measure. But if the vectors are interpreted as lengths the bivector is usually interpreted as an area with the same units, as follows.
- Like the direction of a vector a plane associated with a bivector has a direction, a circulation or a sense of rotation in the plane, which takes two values seen as clockwise and counterclockwise when viewed from viewpoint not in the plane. This is associated with a change of sign in the bivector, that is if the direction is reversed the bivector is negated. Alternately if two bivectors have the same attitude and magnitude but opposite directions then one is the negative of the other.

In three dimensions all bivectors can be generated by the exterior product of two vectors. If the bivector B = a ≠ b then the magnitude of B is

$$|B| = |a||b|\sin\theta,$$

where θ is the angle between the vectors. This is the area of the parallelogram with edges a and b. One interpretation is that the area is swept out by b as it moves along a. The exterior product is antisymmetric, so reversing the order of a and b to make a move along b results in a bivector with the opposite direction that is the negative of the first. The plane of bivector a ≠ b contains both a and b so they are both parallel to the plane.

Bivectors and axial vectors are related by Hodge dual. In a real vector space the Hodge dual relates a subspace to its orthogonal complement, so if a bivector is represented by a plane then the axial vector associated with it is simply the plane's surface normal. The plane has two normals, one on each side, giving the two possible orientations for the plane and bivector.

This relates the cross product to the exterior product. It can also be used to represent physical quantities, like torque and angular momentum. In vector algebra they are usually represented by vectors, perpendicular to the plane of the force, linear momentum or displacement that they are calculated from. But if a bivector is used instead the plane is the plane of the bivector, so is a more natural way to represent the quantities and the way they act. It also unlike the vector representation generalises into other dimensions.

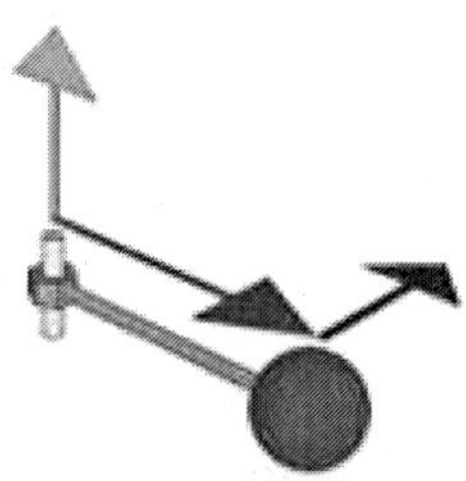

Figure: *Relationship between force F, torque τ, linear momentum p, and angular momentum L.*

The product of two bivectors has a geometric interpretation. For non-zero bivectors A and B the product can be split into symmetric and antisymmetric parts as follows:

$$AB = A \cdot B + A \times B.$$

Like vectors these have magnitudes $|A \cdot B| = |A|\,|B| \cos \theta$ and $|A \times B| = |A|\,|B| \sin \theta$, where θ is the angle between the planes. In three dimensions it is the same as the angle between the normal vectors dual to the planes, and it generalises to some extent in higher dimensions.

Bivectors can be added together as areas. Given two non-zero bivectors B and C in three dimensions it is always possible to find a vector that is contained in both, a say, so the bivectors can be written as exterior products involving a:

$$B = a \wedge b$$
$$C = a \wedge c$$

This can be interpreted geometrically as seen in the diagram: the two areas sum to give a third, with the three areas forming faces of a prism with a, b, c and b + c as edges. This corresponds to the two ways of calculating the area using the distributivity of the exterior product:

$$\begin{aligned} B+C &= a\wedge b + a\wedge c \\ &= a\wedge(b+c). \end{aligned}$$

This only works in three dimensions as it is the only dimension where a vector parallel to both bivectors must exist. In higher dimensions bivectors generally are not associated with a single plane, or if they are (simple bivectors) two bivectors may have no vector in common, and so sum to a non-simple bivector.

Four Dimensions

In four dimensions the basis elements for the space $\wedge^2 \mathbb{R}^4$ of bivectors are (e_{12}, e_{13}, e_{14}, e_{23}, e_{24}, e_{34}), so a general bivector is of the form

$$A = a_{12}e_{12} + a_{13}e_{13} + a_{14}e_{14} + a_{23}e_{23} + a_{24}e_{24} + a_{34}e_{34}.$$

Orthogonality

In four dimensions bivectors are orthogonal to bivectors. That is the dual of a bivector is a bivector, and the space $\wedge^2 \mathbb{R}^4$ is dual to itself in $C\!-\!\ell_4(\mathbb{R})$. Normal vectors are not unique, instead every plane is orthogonal to all the vectors in its dual space. This can be used to partition the bivectors into two 'halves', for example into two sets of three unit bivectors each. There are only four distinct ways to do this, and whenever it's done one vector is in only one of the two halves, for example (e_{12}, e_{13}, e_{14}) and (e_{23}, e_{24}, e_{34}).

Simple Bivectors in 4D

In four dimensions bivectors are generated by the exterior product of vectors in $\mathbb{R}^4$, but with one important difference from $\mathbb{R}^3$ and $\mathbb{R}^2$. In four dimensions not all bivectors are simple. There are bivectors such as $e_{12} + e_{34}$ that cannot be generated by the external product of two vectors. This also means they do not have a real, that is scalar, square. In this case

$$(e_{12}+e_{34})^2 = e_{12}e_{12}+e_{12}e_{34}+e_{34}e_{12}+e_{34}e_{34} = -2+2e_{1234}.$$

The element e_{1234} is the pseudoscalar in $C\ell_4$, distinct from the scalar, so the square is non-scalar.

All bivectors in four dimensions can be generated using at most two exterior products and four vectors. The above bivector can be written as

$$e_{12}+e_{34} = e_1 \wedge e_2 + e_3 \wedge e_4.$$

Alternately every bivector can be written as the sum of two simple bivectors. It is useful to choose two orthogonal bivectors for this, and this is always possible to do. Moreover for a general bivector the choice of simple bivectors is unique, that is there is only one way to decompose into orthogonal bivectors. This is true also for simple bivectors, except one of the orthogonal parts is zero. The exception is when the two orthogonal bivectors have equal magnitudes (as in the above example): in this case the decomposition is not unique.

Rotations in $\mathbb{R}^4$

As in three dimensions bivectors in four dimension generate rotations through the exponential map, and all rotations can be generated this way. As in three dimensions if B is a bivector then the rotor R is $e^{B/2}$ and rotations are generated in the same way:

Figure: *A 3D projection of an tesseract performing an isoclinic rotation.*

The rotations generated are more complex though. They can be categorised as follows:

- *simple* rotations are those that fix a plane in 4D, and rotate by an angle "about" this plane.
- *double* rotations have only one fixed point, the origin, and rotate through two angles about two orthogonal planes. In general the angles are different and the planes are uniquely specified
- *isoclinic* rotations are double rotations where the angles of rotation are equal. In this case the planes about which the rotation is taking place are not unique.

These are generated by bivectors in a straightforward way. Simple rotations are generated by simple bivectors, with the fixed plane the dual or orthogonal to the plane of the bivector.

The rotation can be said to take place about that plane, in the plane of the bivector.

All other bivectors generate double rotations, with the two angles of the rotation equalling the magnitudes of the two simple bivectors the non-simple bivector is composed of. Isoclinic rotations arise when these magnitudes are equal, in which case the decomposition into two simple bivectors is not unique.

Bivectors in general do not commute, but one exception is orthogonal bivectors and exponents of them. So if the bivector $B = B_1 + B_2$, where B_1 and B_2 are orthogonal simple bivectors, is used to generate a rotation it decomposes into two simple rotations that commute as follows:

$$R = e^{\frac{B_1+B_2}{2}} = e^{\frac{B_1}{2}} e^{\frac{B_2}{2}} = e^{\frac{B_2}{2}} e^{\frac{B_1}{2}}$$

It is always possible to do this as all bivectors can be expressed as sums of orthogonal bivectors.

Spacetime Rotations

Spacetime is a mathematical model for our universe used in special relativity. It consists of three space dimensions and one time dimension combined into a single four dimensional space. It is naturally described if using geometric algebra and bivectors, with the Euclidean metric replaced by a Minkowski metric. That is the algebra is identical to that of Euclidean space, except the signature is changed, so

$$e_i^2 = \begin{cases} 1, & i = 1,2,3 \\ -1, & i = 4 \end{cases}$$

(Note the order and indices above are not universal – here e_4 is the time-like dimension). The geometric algebra is $C\ell_{3,1}(\mathbb{R})$, and the subspace of bivectors is $\wedge^2 \mathbb{R}^{3,1}$. The bivectors are of two types. The bivectors e_{23}, e_{31} and e_{12} have negative squares and correspond to the bivectors of the three dimensional subspace corresponding to Euclidean space, $\mathbb{R}^3$. These bivectors generate normal rotations in $\mathbb{R}^3$.

The bivectors e_{14}, e_{24} and e_{34} have positive squares and as planes span a space dimension and the time dimension. These also generate rotations through the exponential map, but instead of trigonometric functions hyperbolic functions are needed, which generates a rotor as follows:

$$e^{\frac{\Omega\theta}{2}} = \cosh\left(\frac{\theta}{2}\right) + \Omega \sinh\left(\frac{\theta}{2}\right).$$

These are Lorentz transformations, expressed in a particularly compact way, using the same algebra as in $\mathbb{R}^3$ and $\mathbb{R}^4$. In general all spacetime rotations are generated from bivectors through the exponential map, that is, a general rotor generated by bivector A is of the form

$$R = e^{\frac{A}{2}}.$$

The set of all rotations in spacetime form the Lorentz group, and from them most of the consequences of special relativity can be deduced. More generally this show how transformations in Euclidean space and spacetime can all be described using the same algebra.

Maxwell's Equations

(Note: in this section traditional 3-vectors are indicated by lines over the symbols and spacetime vector and bivectors by bold symbols, with the vectors J and A exceptionally in uppercase)

Maxwell's equations are used in physics to describe the relationship between electric and magnetic fields. Normally given as four differential equations they have a particularly compact form when the fields are expressed as a spacetime bivector from $\wedge^2 \mathbb{R}^{3,1}$. If the electric and magnetic fields in $\mathbb{R}^3$ are E and B then the *electromagnetic bivector* is

$$F = \frac{1}{c}\overline{E}e_4 + \overline{B}e_{123},$$

where e_4 is again the basis vector for the time-like dimension and c is the speed of light. The quantity Be_{123} is the bivector dual to B in three dimensions, as discussed above, while Ee_4 as a product of orthogonal vectors is also bivector valued. As a whole it is the electromagnetic tensor expressed more compactly as a bivector, and is used as follows. First it is related to the 4-current J, a vector quantity given by

$$J = \overline{j} + c\rho e_4,$$

where j is current density and ρ is charge density. They are related by a differential operator ∂, which is

$$\partial = \nabla - e_4 \frac{1}{c}\frac{\partial}{\partial t}.$$

The operator ∇ is a differential operator in geometric algebra, acting on the space dimensions and given by ∇ M = ∇ ·M + ∇ ≠M. When applied to vectors ∇ ·M is the divergence and ∇ ≠M is the curl but with a bivector rather than vector result, that is dual in three dimensions to the curl. For general quantity M they act as grade lowering and raising differential operators. In particular if M is a scalar then this operator is just the gradient, and it can be thought of as a geometric algebraic del operator.

Together these can be used to give a particularly compact form for Maxwell's equations in a vacuum:

$$\partial F = J.$$

This when decomposed according to geometric algebra, using geometric products which have both grade raising and grade lowering effects, is equivalent to Maxwell's four equations. This is the form in a vacuum, but the general form is only a little more complex. It is also related to the electromagnetic four-potential, a vector A given by

$$A = \overline{\Lambda} + \frac{1}{c}Ve_4,$$

where A is the vector magnetic potential and V is the electric potential. It is related to the electromagnetic bivector as follows

$$\partial A = -F,$$

using the same differential operator.

Higher Dimensions

As has been suggested in earlier sections much of geometric algebra generalises well into higher dimensions. The geometric algebra for the real space $\mathbb{R}^n$ is $C\ell_n(\mathbb{R})$, and the subspace of bivectors is $\wedge^2\mathbb{R}^n$.

The number of simple bivectors needed to form a general bivector rises with the dimension, so for n odd it is $(n - 1) / 2$, for n even it is $n / 2$. So for four and five dimensions only two simple bivectors are needed but three are required for six and seven dimensions. For example in six dimensions with standard basis (e_1, e_2, e_3, e_4, e_5, e_6) the bivector

$$e_{12} + e_{34} + e_{56}$$

is the sum of three simple bivectors but no less. As in four dimensions it is always possible to find orthogonal simple bivectors for this sum.

Rotations in Higher Dimensions

As in three and four dimensions rotors are generated by the exponential map, so

$$e^{\frac{B}{2}}$$

is the rotor generated by bivector B. Simple rotations, that take place in a plane of rotation around a fixed blade of dimension $(n - 2)$ are generated by with simple bivectors, while other bivectors generate more complex rotations which can be described in terms of the simple bivectors they are sums of, each related to a plane of rotation. All bivectors can be expressed as the sum of orthogonal and commutative simple bivectors, so rotations can always be decomposed into a set of commutative rotations about the planes associated with these bivectors. The group of the rotors in n dimensions is the spin group, Spin(n).

One notable feature, related to the number of simple bivectors and so rotation planes, is that in odd dimensions every rotation has a fixed axis - it is misleading to call it an axis of rotation as in higher dimensions rotations are taking place in multiple planes orthogonal to it. This is related to bivectors, as bivectors in odd dimensions decompose into the same number of bivectors as the even dimension below, so have the same number of planes, but one extra dimension. As each plane generates rotations in two dimensions in odd dimensions there must be one dimension, that is an axis, that is not being rotated.

Bivectors are also related to the rotation matrix in n dimensions. As in three dimensions the characteristic equation of the matrix can be solved to find the eigenvalues. In odd dimensions this has one real root, with eigenvector the fixed axis, and in even dimensions it has no real roots, so either all or all but one of the roots are complex conjugate pairs. Each pair is associated with a simple component of the bivector associated with the rotation. In particular the log of each pair is $\pm$ the magnitude, while eigenvectors generated from the roots are parallel to and so can be used to generate the bivector. In general the eigenvalues and bivectors are unique, and the set of eigenvalues gives the full decomposition into simple bivectors; if roots are repeated then the decomposition of the bivector into simple bivectors is not unique.

Chapter 7

Helmholtz Decomposition

Statement of the Theorem

Let F be a vector field on a bounded domain V in R^3, which is twice continuously differentiable. Then F can be decomposed into a curl-free component and a divergence-free component:

$$F = -\nabla\varphi + \nabla \times A,$$

where

$$\varphi(r) = \frac{1}{4\pi}\int_V \frac{\nabla' \cdot F(r')}{|r-r'|} dV' - \frac{1}{4\pi}\int_S \frac{F(r') \cdot dS'}{|r-r'|},$$

$$A(r) = \frac{1}{4\pi}\int_V \frac{\nabla' \times F(r')}{|r-r'|} dV' + \frac{1}{4\pi}\int_S \frac{F(r') \times dS'}{|r-r'|}.$$

If V is R^3 itself (unbounded), and F vanishes sufficiently fast at infinity, then the second component of both scalar and vector potential are zero. That is,

$$\varphi(r) = \frac{1}{4\pi}\int_V \frac{\nabla' \cdot F(r')}{|r-r'|} dV',$$

$$A(r) = \frac{1}{4\pi}\int_V \frac{\nabla' \times F(r')}{|r-r'|} dV'.$$

Fields with Prescribed Divergence and Curl

The term "Helmholtz Theorem" can also refer to the following. Let C be a solenoidal vector field and d a scalar field on R^3 which are sufficiently smooth and which vanish faster than $1/r^2$ at infinity. Then there exists a vector field F such that

$$\nabla \cdot F = d \text{ and } \nabla \times F = C;$$

if additionally the vector field F vanishes as $r \to \infty$, then F is unique. In other words, a vector field can be constructed with both a specified divergence and a specified curl, and if it also vanishes at infinity, it is uniquely specified by its divergence and curl. This theorem is of great importance in electrostatics, since Maxwell's equations for the electric and magnetic fields in the static case are of exactly this type. The proof is by a construction generalizing the one given above: we set

$$F = -\nabla(\mathcal{G}(d)) + \nabla \times (\mathcal{G}(C)),$$

where $\mathcal{G}$ represents the Newtonian potential operator. (When acting on a vector field, such as $\nabla \times$ F, it is defined to act on each component.)

Differential forms

The Hodge decomposition is closely related to the Helmholtz decomposition, generalizing from vector fields on R^3 to differential forms on a Riemannian manifold *M*. Most formulations of the Hodge decomposition require *M* to be compact. Since this is not true of R^3, the Hodge decomposition theorem is not strictly a generalization of the Helmholtz theorem. However, the compactness restriction in the usual formulation of the Hodge decomposition can be replaced by suitable decay assumptions at infinity on the differential forms involved, giving a proper generalization of the Helmholtz theorem.

Weak Formulation

The Helmholtz decomposition can also be generalized by reducing the regularity assumptions (the need for the existence of strong derivatives). Suppose Ω is a bounded, simply-connected, Lipschitz domain. Every square-integrable vector field u " $(L^2(\Omega))^3$ has an orthogonal decomposition:

$$u = \nabla\varphi + \nabla \times A$$

where φ is in the Sobolev space $H^1(\Omega)$ of square-integrable functions on Ω whose partial derivatives defined in the distribution sense are square integrable, and A ∈ *H*(curl,Ω), the Sobolev space of vector fields consisting of square integrable vector fields with square integrable curl.

For a slightly smoother vector field u $\in H(\text{curl},\Omega)$, a similar decomposition holds:

$$u = \nabla\varphi + v$$

where $\varphi \in H^1(\Omega)$ and v $\in (H^1(\Omega))^d$.

Longitudinal and Transverse Fields

A terminology often used in physics refers to the curl-free component of a vector field as the longitudinal component and the divergence-free component as the transverse component. This terminology comes from the following construction: Compute the three-dimensional Fourier transform of the vector field F, which we call $\tilde{F}$. Then decompose this field, at each point k, into two components, one of which points longitudinally, i.e. parallel to k, the other of which points in the transverse direction, i.e. perpendicular to k. So far, we have

$$\tilde{F}(k) = \tilde{F}_l(k) + \tilde{F}_t(k)$$

$$k \cdot \tilde{F}_t(k) = 0.$$

$$k \times \tilde{F}_l(k) = 0.$$

Now we apply an inverse Fourier transform to each of these components. Using properties of Fourier transforms, we derive:

$$F = F_t + F_l$$

$$\nabla \cdot F_t = 0$$

$$\nabla \times F_l = 0$$

so this is indeed the Helmholtz decomposition.

Properties

The following properties can all be derived from the ordinary differentiation rules of calculus. Most importantly, the divergence is a linear operator, i.e.

$$\text{div}(aF + bG) = a\,\text{div}(F) + b\,\text{div}(G)$$

for all vector fields F and G and all real numbers a and b.

There is a product rule of the following type: if φ is a scalar valued function and F is a vector field, then

$$\operatorname{div}(\varphi F) = \operatorname{grad}(\varphi)\cdot F + \varphi \operatorname{div}(F),$$

or in more suggestive notation

$$\nabla\cdot(\varphi F) = (\nabla\varphi)\cdot F + \varphi\,(\nabla\cdot F).$$

Another product rule for the cross product of two vector fields F and G in three dimensions involves the curl and reads as follows:

$$\operatorname{div}(F\times G) = \operatorname{curl}(F)\cdot G - F\cdot\operatorname{curl}(G),$$

or
$$\nabla\cdot(F\times G) = (\nabla\times F)\cdot G - F\cdot(\nabla\times G).$$

The Laplacian of a scalar field is the divergence of the field's gradient:

$$\operatorname{div}(\nabla\varphi) = \Delta\varphi.$$

The divergence of the curl of any vector field (in three dimensions) is equal to zero:

$$\nabla\cdot(\nabla\times F) = 0$$

If a vector field F with zero divergence is defined on a ball in R^3, then there exists some vector field G on the ball with F = curl(G). For regions in R^3 more complicated than this, the latter statement might be false. The degree of *failure* of the truth of the statement, measured by the homology of the chain complex

$$\{\text{scalar fields on } U\}$$
$$\to \{\text{vector fields on } U\}$$
$$\to \{\text{vector fields on } U\}$$
$$\to \{\text{scalar fields on } U\}$$

(where the first map is the gradient, the second is the curl, the third is the divergence) serves as a nice quantification of the complicatedness of the underlying region U. These are the beginnings and main motivations of de Rham cohomology.

Relation with the Exterior Derivative

One can express the divergence as a particular case of the exterior derivative, which takes a 2-form to a 3-form in R^3. Define the current two form

$$j = F_1\,dy\wedge dz + F_2\,dz\wedge dx + F_3\,dx\wedge dy\,.$$

It measures the amount of "stuff" flowing through a surface per unit time in a "stuff fluid" of density $\rho = 1 dx \wedge dy \wedge dz$ moving with local velocity F. Its exterior derivative dj is then given by

$$dj = \left(\frac{\partial F_1}{\partial x} + \frac{\partial F_2}{\partial y} + \frac{\partial F_3}{\partial z} \right) dx \wedge dy \wedge dz = (\nabla \cdot F)\rho$$

Thus, the divergence of the vector field F can be expressed as:

$$\nabla \cdot F = \star d \star F^{\flat}$$

Here the superscript is one of the two musical isomorphisms, and is the Hodge dual. Note however that working with the current two form itself and the exterior derivative is usually easier than working with the vector field and divergence, because unlike the divergence, the exterior derivative commutes with a change of (curvilinear) coordinate system.

Generalizations

The divergence of a vector field can be defined in any number of dimensions. If

$$F = (F_1, F_2, \ldots, F_n),$$

in a Euclidean coordinate system where $x = (x_1, x_2, \ldots, x_n)$ and $dx = (dx_1, dx_2, \ldots, dx_n)$, define

$$\operatorname{div} F = \nabla \cdot F = \frac{\partial F_1}{\partial x_1} + \frac{\partial F_2}{\partial x_2} + \cdots + \frac{\partial F_n}{\partial x_n}.$$

The appropriate expression is more complicated in curvilinear coordinates. For any *n*, the divergence is a linear operator, and it satisfies the "product rule"

$$\nabla \cdot (\varphi F) = (\nabla \varphi) \cdot F + \varphi\, (\nabla \cdot F).$$

for any scalar-valued function φ.

The divergence can be defined on any manifold of dimension n with a volume form (or density) μe.g. a Riemannian or Lorentzian manifold. Generalising the construction of a two form for a vector field on $\mathbb{R}^3$, on such a manifold a vector field X defines a n-1 form $j = i_X \mu$ obtained by contracting X with μ. The divergence is then the function defined by

$$dj = \text{div}(X)\mu$$

Standard formulas for the Lie derivative allow us to reformulate this as

$$\mathcal{L}_X\mu = \text{div}(X)\mu$$

This means that the divergence measures the rate of expansion of a volume element as we let it flow with the vector field.

On a Riemannian or Lorentzian manifold the divergence with respect to the metric volume form can be computed in terms of the Levi Civita connection ∇

$$\text{div}(X) = \nabla \cdot X = X^a_{;a}$$

where the second expression is the contraction of the vector field valued 1-form ∇X with itself and the last expression is the traditional coordinate expression used by physicists.

Divergence can also be generalised to tensors. In Einstein notation, the divergence of a contravariant vector F^μ is given by

$$\nabla \cdot F = \nabla_\mu F^\mu$$

where ∇_μ is the covariant derivative. Equivalently, some authors define the divergence of any mixed tensor by using the "musical notation #":

If T is a (p,q)-tensor (p for the contravariant vector and q for the covariant one), then we define the *divergence of T* to be the $(p,q\text{-}1)$-tensor

$$(\text{div}\, T)(Y_1,\ldots,Y_{q-1}) = \text{trace}(X \mapsto \#(\nabla T)(X,\cdot,Y_1,\ldots,Y_{q-1}))$$

that is we trace the covariant derivative on the *first two* covariant indices.

Curl (Mathematics)

In vector calculus, the curl is a vector operator that describes the infinitesimal rotation of a 3-dimensional vector field. At every point in the field, the curl of that field is represented by a vector. The attributes of this vector (length and direction) characterize the rotation at that point.

The direction of the curl is the axis of rotation, as determined by the right-hand rule, and the magnitude of the curl is the magnitude of rotation. If the vector field represents the flow velocity of a moving fluid, then the curl is the circulation density of the fluid. A vector field whose curl is zero is called irrotational. The curl is a form of differentiation for vector fields. The corresponding form of the fundamental theorem of calculus is Stokes' theorem, which relates the surface integral of the curl of a vector field to the line integral of the vector field around the boundary curve.

The alternative terminology *rotor* or *rotational* and alternative notations rot F and $\nabla \times$F are often used (the former especially in many European countries, the latter, using the del operator and the cross product, is more used in other countries) for *curl* and curl F. Unlike the gradient and divergence, curl does not generalize as simply to other dimensions; some generalizations are possible, but only in three dimensions is the geometrically defined curl of a vector field again a vector field.

This is a similar phenomenon as in the 3 dimensional cross product, and the connection is reflected in the notation $\nabla \times$ for the curl. The name "curl" was first suggested by James Clerk Maxwell in 1871.

Definition

The curl of a vector field F, denoted curl F or $\nabla \times$F, at a point is defined in terms of its projection onto various lines through the point. If is any unit vector, the projection of the curl of F onto is defined to be the limiting value of a closed line integral in a plane orthogonal to as the path used in the integral becomes infinitesimally close to the point, divided by the area enclosed. As such, the curl operator maps C^1 functions from R^3 to R^3 to C^0 functions from R^3 to R^3.

Implicitly, curl is defined by:

$$(\nabla \times F)\cdot\hat{n} \overset{\text{def}}{=} \lim_{A\to 0} \frac{\oint_C F\cdot dr}{|A|}$$

Here $\oint_C F\cdot dr$ is a line integral along the boundary of the area in question, and $|A|$ is the magnitude of the area. If $\hat{\nu}$ is an outward pointing in-plane normal, whereas is the unit-vector perpendicular to the plane, then the orientation of C is chosen so that a vector $\hat{\omega}$ tangent to C is positively oriented if and only if $\{\hat{n}, \hat{v}, \hat{\omega}\}$ forms a positively oriented basis for R^3 (right-hand rule).

The above formula means that the curl of a vector field is defined as the infinitesimal area density of the *circulation* of that field. To this definition fit naturally (i) the Kelvin-Stokes theorem, as a global formula corresponding to the definition, and (ii) the following "easy to memorize" definition of the curl in orthogonal curvilinear coordinates, e.g. in cartesian coordinates, spherical, or cylindrical, or even elliptical or parabolical coordinates:

$$(\text{curl}\, F)_3 = \frac{1}{a_1 a_2} \cdot \left(\frac{\partial(a_2 F_2)}{\partial u_1} - \frac{\partial(a_1 F_1)}{\partial u_2} \right)$$

If (x_1, x_2, x_3) are the Cartesian coordinates and (u_1, u_2, u_3) are the curvilinear coordinates, then $a_i = \sqrt{\sum_{j=1}^{3} \left(\frac{\partial x_j}{\partial u_i} \right)^2}$ is the length of the coordinate vector corresponding to u_i. The remaining two components of curl result from cyclic index-permutation: 3,1,2 → 1,2,3 → 2,3,1.

Intuitive Interpretation

Suppose the vector field describes the velocity field of a fluid flow (maybe a large tank of water or gas) and a small ball is located within the fluid or gas (the centre of the ball being fixed at a certain point). If the ball has a rough surface, the fluid flowing past it will make it rotate.

The rotation axis (oriented according to the right hand rule) points in the direction of the curl of the field at the centre of the ball, and the angular speed of the rotation is half the magnitude of the curl at this point.

Usage

In practice, the above definition is rarely used because in virtually all cases, the curl operator can be applied using some set of curvilinear coordinates, for which simpler representations have been derived. The notation $\nabla \times F$ has its origins in the similarities to the 3 dimensional cross product, and it is useful as a mnemonic in Cartesian coordinates if we take " as a vector differential operator del. Such notation involving operators is common in physics and algebra. If certain coordinate systems are used, for instance, polar-toroidal coordinates (common in plasma physics) using the notation $\nabla \times F$ will yield an incorrect result.

Expanded in Cartesian coordinates, $\nabla \times F$ is, for F composed of $[F_x, F_y, F_z]$:

$$\begin{vmatrix} i & j & k \\ \frac{\partial}{\partial x} & \frac{\partial}{\partial y} & \frac{\partial}{\partial z} \\ F_x & F_y & F_z \end{vmatrix}$$

where i, j, and k are the unit vectors for the *x*-, *y*-, and *z*-axes, respectively. This expands as follows:

$$\left(\frac{\partial F_z}{\partial y} - \frac{\partial F_y}{\partial z}\right)i + \left(\frac{\partial F_x}{\partial z} - \frac{\partial F_z}{\partial x}\right)j + \left(\frac{\partial F_y}{\partial x} - \frac{\partial F_x}{\partial y}\right)k$$

Although expressed in terms of coordinates, the result is invariant under proper rotations of the coordinate axes but the result inverts under reflection.

In a general coordinate system, the curl is given by

$$(\nabla \times F)^k = \epsilon^{k\ell m} \partial_\ell F_m a$$

where ε denotes the Levi-Civita symbol, the metric tensor is used to lower the index on F, and the Einstein summation convention implies that repeated indices are summed over. Equivalently,

$$(\nabla \times F) = e_k \epsilon^{k\ell m} \partial_\ell F_m$$

where e_k are the coordinate vector fields. Equivalently, using the exterior derivative, the curl can be expressed as:

$$\nabla \times F = \left[\star\left(dF^{\flat}\right)\right]^{\sharp}$$

Here and are the musical isomorphisms, and is the Hodge dual. This formula shows how to calculate the curl of F in any coordinate system, and how to extend the curl to any oriented three dimensional Riemannian manifold. Since this depends on a choice of orientation, curl is a chiral operation. In other words, if the orientation is reversed, then the direction of the curl is also reversed.

Identities

Consider the example $\nabla \times [\, v \times F \,]$. Using Cartesian coordinates, it can be shown that

$$\nabla \times (v \times F) = \left[(\nabla \cdot F) + F \cdot \nabla\right] v - \left[(\nabla \cdot v) + v \cdot \nabla\right] F .$$

In the case where the vector field v and " are interchanged:

$$v \times (\nabla \times F) = \nabla_F (v \cdot F) - (v \cdot \nabla) F ,$$

which introduces the Feynman subscript notation "$_F$" which means the subscripted gradient operates only on the factor F.

Another example is $\nabla \times [\, \nabla \times F \,]$. Using Cartesian coordinates, it can be shown that:

$$\nabla \times (\nabla \times F) = \nabla(\nabla \cdot F) - \nabla^2 F ,$$

which can be construed as a special case of the previous example with the substitution $v \to \nabla$.

(*Note:* $\nabla^2 F$ represents, in the case, the vector formed by the individual laplacian of each component of the vector in question) The curl of the gradient of *any* scalar field is always the zero vector:

$$\nabla \times (\nabla \phi) = \vec{0}$$

If $\varphi$is a scalar valued function and F is a vector field, then

$$\nabla \times (\varphi F) = \nabla \varphi \times F + \varphi \nabla \times F$$

Descriptive Examples

- In a vector field describing the linear velocities of each part of a rotating disk, the curl has the same value at all points.

- Of the four Maxwell's equations, two—Faraday's law and Ampère's law—can be compactly expressed using curl. Faraday's law states that the curl of an electric field is equal to the opposite of the time rate of change of the magnetic field, while Ampère's law relates the curl of the magnetic field to the current and rate of change of the electric field.

Generalizations

The vector calculus operations of grad, curl, and div are most easily generalized and understood in the context of differential forms, which involves a number of steps. In a nutshell, they correspond to the derivatives of 0-forms, 1-forms, and 2-forms, respectively. The geometric interpretation of curl as rotation corresponds to identifying bivectors (2-vectors) in 3 dimensions with the special orthogonal Lie algebra of infinitesimal rotations (in coordinates, skew-symmetric 3×3 matrices), while representing rotations by vectors corresponds to identifying 1-vectors (equivalently, 2-vectors) and these all being 3-dimensional spaces.

Differential Form

In the mathematical fields of differential geometry and tensor calculus, differential forms are an approach to multivariable calculus that is independent of coordinates. Differential forms provide a unified approach to defining integrands over curves, surfaces, volumes, and higher dimensional manifolds. The modern notion of differential forms was pioneered by Élie Cartan. It has many applications, especially in geometry, topology and physics.

For instance, the expression $f(x)\,dx$ from one-variable calculus is called a 1-form, and can be integrated over an interval $[a,b]$ in the domain of f:

$$\int_a^b f(x)dx$$

and similarly the expression: $f(x,y,z)\,dx \wedge dy + g(x,y,z)\,dx \wedge dz + h(x,y,z)\,dy \wedge dz$ is a 2-form that has a surface integral over an oriented surface S:

$$\int_S f(x,y,z)dx \wedge dy + g(x,y,z)dx \wedge dz + h(x,y,z)dy \wedge dz.$$

Likewise, a 3-form $f(x, y, z)\, dx \wedge dy \wedge dz$ represents something that can be integrated over a region of space.

The algebra of differential forms is organized in a way that naturally reflects the orientation of the domain of integration. There is an operation d on differential forms known as the exterior derivative that, when acting on a k-form produces a $(k+1)$-form. This operation extends the differential of a function, and the divergence and the curl of a vector field in an appropriate sense that makes the fundamental theorem of calculus, the divergence theorem, Green's theorem, and Stokes' theorem special cases of the same general result, known in this context also as the general Stokes' theorem. In a deeper way, this theorem relates the topology of the domain of integration to the structure of the differential forms themselves; the precise connection is known as De Rham's theorem.

The general setting for the study of differential forms is on a differentiable manifold. Differential 1-forms are naturally dual to vector fields on a manifold, and the pairing between vector fields and 1-forms is extended to arbitrary differential forms by the interior product. The algebra of differential forms along with the exterior derivative defined on it is preserved by the pullback under smooth functions between two manifolds. This feature allows geometrically invariant information to be moved from one space to another via the pullback, provided the information is expressed in terms of differential forms. As a particular example, the change of variables formula for integration becomes a simple statement that an integral is preserved under pullback.

History

Differential forms are part of the field of differential geometry, influenced by linear algebra. Although the notion of a differential is quite old, the initial attempt at an algebraic organization of differential forms is usually credited to Élie Cartan in his 1889 paper .

Concept

Differential forms provide an approach to multivariable calculus that is independent of coordinates.

Let U be an open set in R^n. A differential 0-form ("zero form") is defined to be a smooth function f on U. If v is any vector in R^n, then f has a directional derivative $\partial_v f$, which is another function on U whose value at a point $p \in U$ is the rate of change (at p) of f in the v direction:

$$(\partial_v f)(p) = \frac{d}{dt} f(p+tv)\Big|_{t=0}.$$

(This notion can be extended to the case that v is a vector field on U by evaluating v at the point p in the definition.)

In particular, if $v = e_j$ is the jth coordinate vector then $\partial_v f$ is the partial derivative of f with respect to the jth coordinate function, i.e., $\partial f / \partial x^j$, where $x^1, x^2, \ldots x^n$ are the coordinate functions on U. By their very definition, partial derivatives depend upon the choice of coordinates: if new coordinates $y^1, y^2, \ldots y^n$ are introduced, then

$$\frac{\partial f}{\partial x^j} = \sum_{i=1}^{n} \frac{\partial y^i}{\partial x^j} \frac{\partial f}{\partial y^i}$$

The first idea leading to differential forms is the observation that $\partial_v f(p)$ is a linear function of v:

$$(\partial_{v+w} f)(p) = (\partial_v f)(p) + (\partial_w f)(p)$$

$$(\partial_{cv} f)(p) = c(\partial_v f)(p)$$

for any vectors v, w and any real number c. This linear map from R^n to R is denoted df_p and called the derivative of f at p. Thus $df_p(v) = \partial_v f(p)$.

The object df can be viewed as a function on U, whose value at p is not a real number, but the linear map df_p. This is just the usual Fréchet derivative — an example of a differential 1-form.

Since any vector v is a linear combination $\Sigma\, v^j e_j$ of its components, df is uniquely determined by $df_p(e_j)$ for each j and each $p \in U$, which are just the partial derivatives of f on U.

Thus df provides a way of encoding the partial derivatives of f. It can be decoded by noticing that the coordinates $x^1, x^2, \ldots x^n$ are themselves functions on U, and so define differential 1-forms $dx^1, dx^2, \ldots, dx^n$. Since $\partial x^i / \partial x^j =$ ä$_{ij}$, the Kronecker delta function, it follows that

$$(*) \quad df = \sum_{i=1}^{n} \frac{\partial f}{\partial x^i} dx^i.$$

The meaning of this expression is given by evaluating both sides at an arbitrary point p: on the right hand side, the sum is defined "pointwise", so that

$$df_p = \sum_{i=1}^{n} \frac{\partial f}{\partial x^i}(p)(dx^i)_p.$$

Applying both sides to e_j, the result on each side is the jth partial derivative of f at p. Since p and j were arbitrary, this proves the formula (*).

More generally, for any smooth functions g_i and h_i on U, we define the differential 1-form $\alpha = \Sigma_i\, g_i\, dh_i$ pointwise by

$$\alpha_p = \sum_i g_i(p)(dh_i)_p$$

for each $p \in U$. Any differential 1-form arises this way, and by using (*) it follows that any differential 1-form α on U may be expressed in coordinates as

$$\alpha = \sum_{i=1}^{n} f_i\, dx^i$$

for some smooth functions f_i on U.

The second idea leading to differential forms arises from the following question: given a differential 1-form α on U, when does there exist a function f on U such that $\alpha = df$? The above expansion reduces this question to the search for a function f whose partial derivatives $\partial f / \partial x^i$ are equal to n given functions f_i. For $n>1$, such a function does not always exist: any smooth function f satisfies

$$\frac{\partial f_j}{\partial x^i} - \frac{\partial f_i}{\partial x^j} = 0.$$

so it will be impossible to find such an f unless

$$\frac{\partial f_j}{\partial x^i} - \frac{\partial f_i}{\partial x^j} = 0.$$

for all i and j.

The skew-symmetry of the left hand side in i and j suggests introducing an antisymmetric product on differential 1-forms, the wedge product, so that these equations can be combined into a single condition

$$\sum_{i,j=1}^{n} \frac{\partial f_j}{\partial x^i} dx^i \wedge dx^j = 0$$

where $\quad dx^i \wedge dx^j = -dx^j \wedge dx^i.$

This is an example of a differential 2-form: the exterior derivative $d\alpha$ of $\alpha = \Sigma_{j=1}^{n} f_j\, dx^j$ is given by

$$d\alpha = \sum_{j=1}^{n} df_j \wedge dx^j = \sum_{i,j=1}^{n} \frac{\partial f_j}{\partial x^i} dx^i \wedge dx^j.$$

To summarize: $d\alpha = 0$ is a necessary condition for the existence of a function f with $\alpha = df$.

Differential 0-forms, 1-forms, and 2-forms are special cases of differential forms. For each k, there is a space of differential k-forms, which can be expressed in terms of the coordinates as

$$\sum_{i_1, i_2 \ldots i_k = 1}^{n} f_{i_1 i_2 \ldots i_k} dx^{i_1} \wedge dx^{i_2} \wedge \cdots \wedge dx^{i_k}$$

for a collection of functions $f_{i1i2 \ldots ik}$. (Of course, as assumed below, one can restrict the sum to the case $i_1 < i_2 < \cdots < i_{n-1} < i_n$.)

Differential forms can be multiplied together using the wedge product, and for any differential k-form α, there is a differential $(k + 1)$-form $d\alpha$ called the exterior derivative of α.

Differential forms, the wedge product and the exterior derivative are independent of a choice of coordinates. Consequently they may be defined on any smooth manifold M. One way to do this is cover M with coordinate charts and define a differential k-form on M to be a family of differential k-forms on each chart which agree on the overlaps. However, there are more intrinsic definitions which make the independence of coordinates manifest.

Intrinsic Definitions

Let M be a smooth manifold. A differential form of degree k is a smooth section of the kth exterior power of the cotangent bundle of M. At any point $p \in M$, a k-form β defines an alternating multilinear map

$$\beta_p : T_pM \times \cdots \times T_pM \to \mathbb{R}$$

(with k factors of T_pM in the product), where T_pM is the tangent space to M at p. Equivalently, β is a totally antisymmetric covariant tensor field of rank k.

The set of all differential k-forms on a manifold M is a vector space, often denoted $\Omega^k(M)$.

For example, a differential 1-form α assigns to each point $p \in M$ a linear functional α_p on T_pM. In the presence of an inner product on T_pM (induced by a Riemannian metric on M), α_p may be represented as the inner product with a tangent vector X_p. Differential 1-forms are sometimes called covariant vector fields, covector fields, or "dual vector fields", particularly within physics.

Operations

There are several operations on differential forms: the wedge product of two differential forms, the exterior derivative of a single differential form, the interior product of a differential form and a vector field, and the Lie derivative of a differential form with respect to a vector field.

Wedge Product

The wedge product of a k-form α and an l-form β is a $(k + l)$-form denoted $\alpha \wedge \beta$. For example, if $k = l = 1$, then $\alpha \wedge \beta$ is the 2-form whose value at a point p is the alternating bilinear form defined by

$$(\alpha \wedge \beta)_p(v, w) = \alpha_p(v)\beta_p(w) - \alpha_p(w)\beta_p(v)$$

for v, $w \in T_pM$. (In an alternative convention, the right hand side is divided by two in this formula.)

The wedge product is bilinear: for instance, if α, β, and γ are any differential forms, then

$$\alpha \wedge (\beta + \gamma) = \alpha \wedge \beta + \alpha \wedge \gamma.$$

It is *skew commutative* (also known as *graded commutative*), meaning that it satisfies a variant of anticommutativity that depends on the degrees of the forms: if α is a k-form and β is an l-form, then

$$\alpha \wedge \beta = (-1)^{kl} \beta \wedge \alpha.$$

Riemannian Manifold

On a Riemannian manifold, or more generally a pseudo-Riemannian manifold, vector fields and covector field can be identified (the metric is a fibre-wise isomorphism of the tangent space and the cotangent space), and additional operations can thus be defined, such as the Hodge star operator $* : \Omega^k(M) \tilde{\rightarrow} \Omega^{n-k}(M)$ and codifferential $\delta : \Omega^k(M) \rightarrow \Omega^{k-1}(M)$, (degree -1) which is adjoint to the exterior differential d.

Vector Field Structures

On a pseudo-Riemannian manifold, 1-forms can be identified with vector fields; vector fields have additional distinct algebraic structures, which are listed here for context and to avoid confusion.

Firstly, each (co)tangent space generates a Clifford algebra, where the product of a (co)vector with itself is given by the value of a quadratic form - in this case, the natural one induced by the metric.

This algebra is *distinct* from the exterior algebra of differential forms, which can be viewed as a Clifford algebra where the quadratic form vanishes (since the exterior product of any vector with itself is zero). Clifford algebras are thus non-anti-commutative ("quantum") deformations of the exterior algebra. They are studied in geometric algebra.

Another alternative is to consider vector fields as derivations, and consider the (noncommutative) algebra of differential operators they generate, which is the Weyl algebra, and is a noncommutative ("quantum") deformation of the *symmetric* algebra in the vector fields.

Exterior Differential Complex

One important property of the exterior derivative is that $d^2 = 0$. This means that the exterior derivative defines a cochain complex:

$$0 \rightarrow \Omega^0(M) \xrightarrow{d} \Omega^1(M) \xrightarrow{d} \Omega^2(M) \xrightarrow{d} \Omega^3(M) \rightarrow \cdots \rightarrow \Omega^n(M) \rightarrow 0.$$

By the Poincaré lemma, this complex is locally exact except at Ω^0(M). Its cohomology is the de Rham cohomology of M.

Pullback

One of the main reasons the cotangent bundle rather than the tangent bundle is used in the construction of the exterior complex is that differential forms are capable of being pulled back by smooth maps, while vector fields cannot be pushed forward by smooth maps unless the map is, say, a diffeomorphism.

The existence of pullback homomorphisms in de Rham cohomology depends on the pullback of differential forms. Differential forms can be moved from one manifold to another using a smooth map. If $f : M \to N$ is smooth and Ω is a smooth k-form on N, then there is a differential form f^*ù on M, called the pullback of ω, which captures the behaviour of ω as seen relative to f.

To define the pullback, recall that the differential of f is a map $f_* : TM \to TN$. Fix a differential k-form ω on N. For a point p of M and tangent vectors $v_1, \ldots, v_k$ to M at p, the pullback of ω is defined by the formula

$$(f^*\omega)_p(v_1,\ldots,v_k) = \omega_{f(p)}(f_*v_1,\ldots,f_*v_k).$$

More abstractly, if ω is viewed as a section of the cotangent bundle T^*N of N, then $f^*\omega$ is the section of T^*M defined as the composite map

$$M \xrightarrow{f} N \xrightarrow{\omega} T^*N \xrightarrow{(Df)^*} T^*M.$$

Pullback respects all of the basic operations on forms:

$$f^*(\omega+\eta) = f^*\omega + f^*\eta,$$

$$f^*(\omega\wedge\eta) = f^*\omega \wedge f^*\eta,$$

$$f^*(d\omega) - d(f^*\omega).$$

The pullback of a form can also be written in coordinates. Assume that $x_1, \ldots, x_m$ $\omega_{i_1\cdots i_k}$ are coordinates on M, that $y_1, \ldots, y_n$ are coordinates on N, and that these coordinate systems are related by the formulas $y_i = f_i(x_1, \ldots, x_m)$ for all i. Then, locally on N, ω can be written as

$$\omega = \sum_{i_1<\cdots<i_k} \omega_{i_1\cdots i_k} dy_{i_1} \wedge\cdots\wedge dy_{i_k},$$

where, for each choice of $i_1, \ldots, i_k$, is a real-valued function of $y_1, \ldots, y_n$. Using the linearity of pullback and its compatibility with wedge product, the pullback of ω has the formula

$$f^*\omega = \sum_{i_1<\cdots<i_k} (\omega_{i_1\cdots i_k} \circ f) df_{i_1} \wedge \cdots \wedge df_{i_n}.$$

Each exterior derivative df_i can be expanded in terms of $dx_1, \ldots, dx_m$. The resulting k-form can be written using Jacobian matrices:

$$f^*\omega = \sum_{i_1<\cdots<i_k} \sum_{j_1<\cdots<j_k} (\omega_{i_1\cdots i_k} \circ f) \frac{\partial(f_{i_1}, \ldots, f_{i_k})}{\partial(x_{j_1}, \ldots, x_{j_k})} dx_{j_1} \wedge \cdots \wedge dx_{j_k}.$$

Relation with Measures

On a *general* differentiable manifold (without additional structure), differential forms *cannot* be integrated over subsets of the manifold; this distinction is key to the distinction between differential forms, which are integrated over chains, and measures, which are integrated over subsets. The simplest example is attempting to integrate the 1-form dx over the interval [0,1].

Assuming the usual distance (and thus measure) on the real line, this integral is either 1 or -1, depending on *orientation:* $\int_0^1 dx = 1$, while $\int_1^0 dx = -\int_0^1 dx = -1$. By contrast, the integral of the *measure* dx on the interval is unambiguously 1 (formally, the integral of the constant function 1 with respect to this measure is 1). Similarly, under a change of coordinates a differential n-form changes by the Jacobian determinant J, while a measure changes by the *absolute value* of the Jacobian determinant, $|J|$, which further reflects the issue of orientation. For example, under the map $x \mapsto -x$ on the line, the differential form pulls back to orientation has reversed; while the Lebesgue measure, also denoted dx pulls back to dx; it does not change.

In the presence of the additional data of an *orientation,* it is possible to integrate n-forms (top-dimensional forms) over the entire manifold or over compact subsets; integration over the entire manifold corresponds to integrating the form over the fundamental class of the manifold, $[M]$. Formally, in the presence of an orientation, one may identify n-forms with densities on a manifold; densities in turn define a measure, and thus can be integrated (Folland 1999, Section 11.4). On an orientable but not oriented manifold, there are two choices of orientation; either choice allows one to integrate n-forms over compact subsets, with the two choices differing by a sign. On non-orientable manifold, n-forms and densities cannot be identified - notably, any top-dimensional form must vanish (there are no volume forms on non-orientable manifolds), but there are non-vanishing densities - thus while one can integrate densities over compact subsets, one cannot integrate n-forms.

There is in general no meaningful way to integrate k-forms over subsets for $k < n$ because there is no consistent way to orient k-dimensional subsets; geometrically, a k-dimensional subset can be turned around in place, reversing any orientation but yielding the same subset. Compare the Gram determinant of a set of k vectors in an n-dimensional space, which, unlike the determinant of n vectors, is always positive, corresponding to a squared number. On a Riemannian manifold, one may define a k-dimensional Hausdorff measure for any k (integer or real), which may be integrated over k-dimensional subsets of the manifold. A function times this Hausdorff measure can then be integrated over k-dimensional subsets, providing a measure-theoretic analog to integration of k-forms. The n-dimensional Hausdorff measure yields a density, as above.

Applications in Physics

Differential forms arise in some important physical contexts. For example, in Maxwell's theory of electromagnetism, the Faraday 2-form, or electromagnetic field strength, is

$$F = \frac{1}{2} f_{ab}\, dx^a \wedge dx^b,$$

where the are formed from the electromagnetic fields $\vec{E}$ and $\vec{B}$ $f_{1,2} = E_z / c$, $f_{2,3} = -B_z$, e.g. , or equivalent definitions.

This form is a special case of the curvature form on the U(1) principal fibre bundle on which both electromagnetism and general gauge theories may be described. The connection form for the principal bundle is the vector potential, typically denoted by A, when represented in some gauge. One then has

$$F = dA.$$

The current 3-form is

$$J = \frac{1}{6} j^a \epsilon_{abcd}\, dx^b \wedge dx^c \wedge dx^d,$$

where j^a are the four components of the current-density. (Here it is a matter of convention, to write F_{ab} instead of f_{ab}, i.e. to use capital letters, and to write J^a instead of j^a.

However, it should be noted that the vector rsp. tensor components and the above-mentioned forms have different physical dimensions. Moreover, one should remember that by decision of an international commission of the IUPAP, the magnetic polarization vector is called $\vec{J}$ since several decades, and by some publishers J, i.e. the same name is used for totally different quantities.)

Using the above-mentioned definitions, Maxwell's equations can be written very compactly in geometrized units as

$$dF = 0$$

$$d^* F = J$$

where denotes the Hodge star operator. Similar considerations describe the geometry of gauge theories in general.

The 2-form $*F$, which is dual to the Faraday form, is also called Maxwell 2-form.

Electromagnetism is an example of a U(1) gauge theory. Here the Lie group is U(1), the one-dimensional unitary group, which is in particular abelian. There are gauge theories, such as Yang-Mills theory, in which the Lie group is not abelian. In that case, one gets relations which are similar to those described here. The analog of the field F in such theories is the curvature form of the connection, which is represented in a gauge by a Lie algebra-valued one-form A. The Yang-Mills field F is then defined by

$$F = dA + A \wedge A.$$

In the abelian case, such as electromagnetism, $A \wedge A = 0$, but this does not hold in general. Likewise the field equations are modified by additional terms involving wedge products of A and F, owing to the structure equations of the gauge group.

Applications in Geometric Measure Theory

Numerous minimality results for complex analytic manifolds are based on the Wirtinger inequality for 2-forms.

A succinct proof may be found in Herbert Federer's classic text Geometric Measure Theory. The Wirtinger inequality is also a key ingredient in Gromov's inequality for complex projective space in systolic geometry.

Curl Geometrically

2-vectors correspond to the exterior power $\Lambda^2 V$; in the presence of an inner product, in coordinates these are the skew-symmetric matrices, which are geometrically considered as the special orthogonal Lie algebra $so(V)$ of infinitesimal rotations.

This has $\binom{n}{2} = \frac{1}{2}n(n-1)$ dimensions, and allows one to interpret the differential of a 1-vector field as its infinitesimal rotations. Only in 3 dimensions (or trivially in 0 dimensions) is $n = \frac{1}{2}n(n-1)$, which is the most elegant and common case. In 2 dimensions the curl of a vector field is not a vector field but a function, as 2-dimensional rotations are given by an angle (a scalar - an orientation is required to choose whether one counts clockwise or counterclockwise rotations as positive); note that this is not the div, but is rather perpendicular to it.

In 3 dimensions the curl of a vector field is a vector field as is familiar (in 1 and 0 dimensions the curl of a vector field is 0, because there are no non-trivial 2-vectors), while in 4 dimensions the curl of a vector field is, geometrically, at each point an element of the 6 dimensional Lie algebra

Note also that the curl of a 3-dimensional vector field which only depends on 2 coordinates (say x, y) is simply a vertical vector field (in the z direction) whose magnitude is the curl of the 2-dimensional vector field, as in the examples on this page.

Considering curl as a 2-vector field (an antisymmetric 2-tensor) has been used to generalize vector calculus and associated physics to higher dimensions.

Laplace Operator

In mathematics the Laplace operator or Laplacian is a differential operator given by the divergence of the gradient of a function on Euclidean space. It is usually denoted by the symbols $\nabla\cdot\nabla, \nabla^2$ or Δ. The Laplacian $\Delta f(p)$ of a function f at a point p, up to a constant depending on the dimension, is the rate at which the average value of f over spheres centred at p, deviates from $f(p)$ as the radius of the sphere grows. In a Cartesian coordinate system, the Laplacian is given by sum of second partial derivatives of the function with respect to each independent variable. In other coordinate systems such as cylindrical and spherical coordinates, the Laplacian also has a useful form. The Laplace operator is named after the French mathematician Pierre-Simon de Laplace (1749–1827), who first applied the operator to the study of celestial mechanics, where the operator gives a constant multiple of the mass density when it is applied to a given gravitational potential. Solutions of the equation $\Delta f = 0$, now called Laplace's equation, are the so-called harmonic functions, and represent the possible gravitational fields in free space.

The Laplacian occurs in differential equations that describe many physical phenomena, such as electric and gravitational potentials, the diffusion equation for heat and fluid flow, wave propagation, and quantum mechanics. The Laplacian represents the flux density of the gradient flow of a function. For instance, the net rate at which a chemical dissolved in a fluid moves toward or away from some point is proportional to the Laplacian of the chemical concentration at that point; expressed symbolically, the resulting equation is the diffusion equation.

For these reasons, it is extensively used in the sciences for modelling all kinds of physical phenomena. The Laplacian is the simplest elliptic operator, and is at the core of Hodge theory as well as the results of de Rham cohomology. In image processing and computer vision, the Laplacian operator has been used for various tasks such as blob and edge detection.

Density Associated to a Potential

If φ denotes the electrostatic potential associated to a charge distribution q, then the charge distribution itself is given by the Laplacian of φ:

$$q = \Delta\varphi.$$

This is a consequence of Gauss's law. Indeed, if V is any smooth region, then by Gauss's law the flux of the electrostatic field E is equal to the charge enclosed (in appropriate units):

$$\int_{\partial V} E\cdot n = \int_{\partial V} \nabla\varphi\cdot n = \int_V q\,dV,$$

where the first equality uses the fact that the electrostatic field is the gradient of the electrostatic potential. The divergence theorem now gives

$$\int_V \Delta\varphi\,dV = \int_V q\,dV,$$

and since this holds for all regions V, (1) follows.

The same approach implies that the Laplacian of the gravitational potential is the mass distribution. Often the charge (or mass) distribution are given, and the associated potential is unknown. Finding the potential function subject to suitable boundary conditions is equivalent to solving Poisson's equation.

Coordinate Expressions

Two Dimensions

The Laplace operator in two dimensions is given by

$$\Delta f = \frac{\partial^2 f}{\partial x^2} + \frac{\partial^2 f}{\partial y^2}$$

where x and y are the standard Cartesian coordinates of the xy-plane. In polar coordinates,

$$\begin{aligned}\Delta f &= \frac{1}{r}\frac{\partial}{\partial r}\left(r\frac{\partial f}{\partial r}\right) + \frac{1}{r^2}\frac{\partial^2 f}{\partial \phi^2} \\ &= \frac{1}{r}\frac{\partial f}{\partial r} + \frac{\partial^2 f}{\partial r^2} + \frac{1}{r^2}\frac{\partial^2 f}{\partial \phi^2}.\end{aligned}$$

Three Dimensions

In three dimensions, it is common to work with the Laplacian in a variety of different coordinate systems. In Cartesian coordinates,

$$\Delta f = \frac{\partial^2 f}{\partial x^2} + \frac{\partial^2 f}{\partial y^2} + \frac{\partial^2 f}{\partial z^2}.$$

In cylindrical coordinates,

$$\Delta f = \frac{1}{\rho}\frac{\partial}{\partial \rho}\left(\rho \frac{\partial f}{\partial \rho}\right) + \frac{1}{\rho^2}\frac{\partial^2 f}{\partial \phi^2} + \frac{\partial^2 f}{\partial z^2}.$$

In spherical coordinates:

$$\Delta f = \frac{1}{r^2}\frac{\partial}{\partial r}\left(r^2 \frac{\partial f}{\partial r}\right) + \frac{1}{r^2 \sin\theta}\frac{\partial}{\partial \theta}\left(\sin\theta \frac{\partial f}{\partial \theta}\right) + \frac{1}{r^2 \sin^2\theta}\frac{\partial^2 f}{\partial \phi^2}.$$

(here φ represents the azimuthal angle and θ the zenith angle or co-latitude).

In general curvilinear coordinates (ξ^1, ξ^2, ξ^3):

$$\nabla^2 = \nabla\xi^m \cdot \nabla\xi^n \frac{\partial^2}{\partial\xi^m \partial\xi^n} + \nabla^2\xi^m \frac{\partial}{\partial\xi^m},$$

where summation over the repeated indices is implied.

N *Dimensions*

In spherical coordinates in N dimensions, with the parametrization $x = r\theta \in \mathrm{R}^N$ with r representing a positive real radius and θ an element of the unit sphere S^{N-1},

$$\Delta f = \frac{\partial^2 f}{\partial r^2} + \frac{N-1}{r}\frac{\partial f}{\partial r} + \frac{1}{r^2}\Delta_{S^{N-1}} f$$

where $\Delta_{S^{N-1}}$ is the Laplace–Beltrami operator on the (N-1)-sphere, known as the spherical Laplacian. The two radial terms can be equivalently rewritten as

$$\frac{1}{r^{N-1}}\frac{\partial}{\partial r}\left(r^{N-1}\frac{\partial f}{\partial r}\right).$$

As a consequence, the spherical Laplacian of a function defined on $S^{N-1} \subset \mathrm{R}^N$ can be computed as the ordinary Laplacian of the function extended to $\mathrm{R}^N\backslash\{0\}$ so that it is constant along rays, i.e., homogeneous of degree zero.

Spectral Theory

The spectrum of the Laplace operator consists of all eigenvalues λ for which there is a corresponding eigenfunction f with

$$-\Delta f = \lambda f.$$

If Ω is a bounded domain in R^n then the eigenfunctions of the Laplacian are an orthonormal basis for the Hilbert space $L^2(\Omega)$. This result essentially follows from the spectral theorem on compact self-adjoint operators, applied to the inverse of the Laplacian (which is compact, by the Poincaré inequality and Kondrakov embedding theorem). It can also be shown that the eigenfunctions are infinitely differentiable functions.

More generally, these results hold for the Laplace–Beltrami operator on any compact Riemannian manifold with boundary, or indeed for the Dirichlet eigenvalue problem of any elliptic operator with smooth coefficients on a bounded domain. When Ω is the n-sphere, the eigenfunctions of the Laplacian are the well-known spherical harmonics.

Laplace–Beltrami Operator

In differential geometry, the Laplace operator, named after Pierre-Simon Laplace, can be generalized to operate on functions defined on surfaces in Euclidean space and, more generally, on Riemannian and pseudo-Riemannian manifolds. This more general operator goes by the name Laplace–Beltrami operator, after Laplace and Eugenio Beltrami. Like the Laplacian, the Laplace–Beltrami operator is defined as the divergence of the gradient, and is a linear operator taking functions into functions. The operator can be extended to operate on tensors as the divergence of the covariant derivative. Alternatively, the operator can be generalized to operate on differential forms using the divergence and exterior derivative. The resulting operator is called the Laplace–de Rham operator (named after Georges de Rham). The Laplace–Beltrami operator, like the Laplacian, is the divergence of the gradient:

$$\Delta f = \operatorname{div} \operatorname{grad} f.$$

An explicit formula in local coordinates is possible.

Suppose first that M is an oriented Riemannian manifold. The orientation allows one to specify a definite volume form on M, given in an oriented coordinate system x^i by

$$\operatorname{vol}_n := \sqrt{|g|}\, dx^1 \wedge \ldots \wedge dx^n$$

where the dx^i are the 1-forms forming the dual basis to the basis vectors

$$\partial_i := \frac{\partial}{\partial x^i}$$

and is the wedge product. Here $|g| := |\det(g_{ij})|$ is the absolute value of the determinant of the metric tensor g_{ij}. The divergence div X of a vector field X on the manifold is then defined as the scalar function with the property

$$(\text{div}X)\ \text{vol}_n := L_X \text{vol}_n$$

where L_X is the Lie derivative along the vector field X. In local coordinates, one obtains

$$\text{div}X = \frac{1}{\sqrt{|g|}} \partial_i \left(\sqrt{|g|} X^i \right)$$

where the Einstein notation is implied, so that the repeated index i is summed over. The gradient of a scalar function f is the vector field grad f that may be defined through the inner product $\langle \cdot,\cdot \rangle$ on the manifold, as

$$\langle \text{grad} f(x), v_x \rangle = df(x)(v_x)$$

for all vectors v_x anchored at point x in the tangent space $T_x M$ of the manifold at point x. Here, df is the exterior derivative of the function f; it is a 1-form taking argument v_x. In local coordinates, one has

$$(\text{grad} f)^i = \partial^i f = g^{ij} \partial_j f$$

where g^{ij} are the components of the inverse of the metric tensor, so that $g^{ij} g_{jk} = \delta^i_k$ with δ^i_k the Kronecker delta.

Combining the definitions of the gradient and divergence, the formula for the Laplace–Beltrami operator Δ applied to a scalar function f is, in local coordinates

$$\Delta f = \text{div grad } f = \frac{1}{\sqrt{|g|}} \partial_i \left(\sqrt{|g|} g^{ij} \partial_j f \right).$$

If M is not oriented, then the above calculation carries through exactly as presented, except that the volume form must instead be replaced by a volume element (a density rather than a form). Neither the gradient nor the divergence actually depends on the choice of orientation, and so the Laplace–Beltrami operator itself does not depend on this additional structure.

Formal Self-Adjointness

The exterior derivative d and "div are formal adjoints, in the sense that for f a compactly supported function

$$\int_M df(X)\,\mathrm{vol}_n = -\int_M f\mathrm{div}X\,\mathrm{vol}_n$$

where the last equality is an application of Stokes' theorem. Dualizing gives

$$\int_M f\,\Delta h\mathrm{vol}_n = -\int_M \langle df, dh\rangle \mathrm{vol}_n$$

for all compactly supported functions f and h. Conversely, (2) characterizes Δ completely, in the sense that it is the only operator with this property.

As a consequence, the Laplace–Beltrami operator is negative and formally self-adjoint, meaning that for compactly supported functions f and h,

$$\int_M f\,\Delta h\,\mathrm{vol}_n = -\int_M \langle df, dh\rangle\,\mathrm{vol}_n = \int_M h\Delta f\,\mathrm{vol}_n.$$

Because the Laplace–Beltrami operator, as defined in this manner, is negative rather than positive, often it is defined with the opposite sign.

Tensor Laplacian

The Laplace–Beltrami operator can be written using the trace of the iterated covariant derivative associated to the Levi-Civita connection. From this perspective, let X_i be a basis of tangent vector fields (not necessarily induced by a coordinate system). Then the Hessian of a function f is the symmetric 2-tensor whose components are given by

$$H(f)_{ij} = H_f(X_i, X_j) = \nabla_{X_i}\nabla_{X_j} f - \nabla_{\nabla_{X_i}X_j} f$$

This is easily seen to transform tensorially, since it is linear in each of the arguments X_i, X_j. The Laplace–Beltrami operator is then the trace of the Hessian with respect to the metric:

$$\Delta f = \sum_{ij} g^{ij} H(f)_{ij}.$$

In abstract indices, the operator is often written

$$\Delta f = \nabla^a \nabla_a f$$

provided it is understood implicitly that this trace is in fact the trace of the Hessian *tensor*.

Because the covariant derivative extends canonically to arbitrary tensors, the Laplace–Beltrami operator defined on a tensor T by

$$\Delta T = g^{ij}\left(\nabla_{X_i}\nabla_{X_j}T - \nabla_{\nabla_{X_i}X_j}T\right)$$

is well-defined.

Laplace–de Rham Operator

More generally, one can define a Laplacian differential operator on sections of the bundle of differential forms on a pseudo-Riemannian manifold. On a Riemannian manifold it is an elliptic operator, while on a Lorentzian manifold it is hyperbolic. The Laplace–de Rham operator is defined by

$$\Delta = \mathrm{d}\delta + \delta\mathrm{d} = (\mathrm{d}+\delta)^2,$$

where d is the exterior derivative or differential and δ is the codifferential, acting as $(-1)^{kn+n+1} * d *$ on k-forms where $*$ is the Hodge star. When computing Δf for a scalar function f, we have $\delta f = 0$, so that

$$\Delta f = \delta df.$$

Up to an overall sign, The Laplace–de Rham operator is equivalent to the previous definition of the Laplace–Beltrami operator when acting on a scalar function. On functions, the Laplace–de Rham operator is actually the negative of the Laplace–Beltrami operator, as the conventional normalization of the codifferential assures that the Laplace–de Rham operator is (formally) positive definite, whereas the Laplace–Beltrami operator is typically negative.

The sign is a pure convention, however, and both are common in the literature. The Laplace–de Rham operator differs more significantly from the tensor Laplacian restricted to act on skew-symmetric tensors. Apart from the incidental sign, the two operators differ by a Weitzenböck identity that explicitly involves the Ricci curvature tensor.

Euclidean Space

In the usual (orthonormal) Cartesian coordinates x^i on Euclidean space, the metric is reduced to the Kronecker delta, and one therefore has $|g|=1$. Consequently, in this case

$$\Delta f = \frac{1}{\sqrt{|g|}}\partial_i\sqrt{|g|}\partial^i f = \partial_i\partial^i f$$

which is the ordinary Laplacian. In curvilinear coordinates, such as spherical or cylindrical coordinates, one obtains alternative expressions.

Similarly, the Laplace–Beltrami operator corresponding to the Minkowski metric with signature ("+++) is the D'Alembertian.

Spherical Laplacian

The spherical Laplacian is the Laplace–Beltrami operator on the $(n - 1)$-sphere with its canonical metric of constant sectional curvature 1. It is convenient to regard the sphere as isometrically embedded into R^n as the unit sphere centred at the origin. Then for a function f on S^{n-1}, the spherical Laplacian is defined by

$$\Delta_{S^{n-1}} f(x) = \Delta f(x/|x|)$$

where $f(x/|x|)$ is the degree zero homogeneous extension of the function f to $R^n - \{0\}$, and Δ is the Laplacian of the ambient Euclidean space. Concretely, this is implied by the well-known formula for the Euclidean Laplacian in spherical polar coordinates:

$$\Delta f = r^{1-n}\frac{\partial}{\partial r}\left(r^{n-1}\frac{\partial f}{\partial r}\right) + r^{-2}\Delta_{S^{n-1}} f.$$

More generally, one can formulate a similar trick using the normal bundle to define the Laplace–Beltrami operator of any Riemannian manifold isometrically embedded as a hypersurface of Euclidean space.

One can also give an intrinsic description of the Laplace–Beltrami operator on the sphere in a normal coordinate system. Let (t, ∂) be spherical coordinates on the sphere with respect to a particular point p of the sphere (the "north pole"), that is geodesic polar coordinates with respect to p. Here t represents the latitude measurement along a unit speed geodesic from p, and ∂ a parameter representing the choice of direction of the geodesic in S^{n-1}. Then the spherical Laplacian has the form:

$$\Delta_{S^{n-1}} f(t,\xi) = \sin^{2-n} t \frac{\partial}{\partial t}\left(\sin^{n-2} t \frac{\partial f}{\partial t} \right) + \sin^{-2} t \Delta_\xi f$$

where Δ_ξ is the Laplace–Beltrami operator on the ordinary unit $(n - 2)$-sphere.

D'Alembertian

The Laplacian can be generalized in certain ways to non-Euclidean spaces, where it may be elliptic, hyperbolic, or ultrahyperbolic.

In the Minkowski space the Laplace–Beltrami operator becomes the d'Alembert operator or d'Alembertian:

$$\Box = \frac{1}{c^2}\frac{\partial^2}{\partial t^2} - \frac{\partial^2}{\partial x^2} - \frac{\partial^2}{\partial y^2} - \frac{\partial^2}{\partial z^2}.$$

It is the generalisation of the Laplace operator in the sense that it is the differential operator which is invariant under the isometry group of the underlying space and it reduces to the Laplace operator if restricted to time independent functions. Note that the overall sign of the metric here is chosen such that the spatial parts of the operator admit a negative sign, which is the usual convention in high energy particle physics. The D'Alembert operator is also known as the wave operator, because it is the differential operator appearing in the wave equations and it is also part of the Klein–Gordon equation, which reduces to the wave equation in the massless case. The additional factor of c in the metric is needed in physics if space and time are measured in different units; a similar factor would be required if, for example, the x direction were measured in metres while the y direction were measured in centimetres. Indeed, theoretical physicists usually work in units such that $c=1$ in order to simplify the equation.

Gradient Theorem

The gradient theorem, also known as the fundamental theorem of calculus for line integrals, says that a line integral through a gradient field (any conservative vector field can be expressed as the gradient of a scalar field) can be evaluated by evaluating the original scalar field at the endpoints of the curve:

$$\phi(q)-\phi(p)=\int_{\gamma[p,q]}\nabla\phi(r)\cdot dr.$$

It is a generalization of the fundamental theorem of calculus to any curve in a plane or space (generally n-dimensional) rather than just the real line.

The gradient theorem implies that line integrals through irrotational vector fields are path independent. In physics this theorem is one of the ways of defining a "conservative" force. By placing ϕ as potential, $\nabla\phi$ is a conservative field. Work done by conservative forces does not depend on the path followed by the object, but only the end points, as the above equation shows.

Proof

If is a differentiable function from some subset $U \subseteq \mathbb{R}^n$ to R, and if r is a continuous function from some closed interval $[a,b]$ to U, then by the multivariate chain rule, the equality

$$\frac{d}{dt}\phi(r(t))=\nabla\phi(r(t))\cdot r'(t)$$

holds for all $t \in [a,b]$, where the denotes the usual inner product.

Therefore, r if parametrizes the (continuous) curve $\gamma \subset U$ with endpoints p,q (oriented in the direction from P to q) for $t \in [a,b]$, then

$$\begin{aligned}\int_{\gamma}\nabla\phi(x)\cdot dx &= \int_a^b\nabla\phi(r(t))\cdot r'(t)dt\\ &= \int_a^b\frac{d}{dt}\phi(r(t))dt=\phi(r(b))-\phi(r(a))=\phi(q)-\phi(p)\end{aligned}$$

where we have used the definition of the line integral in the first equality, and the fundamental theorem of calculus in the third equality.

Generalizations

In the language of differential forms and exterior derivatives, the gradient theorem states that

$$\int_{\partial\gamma}\phi=\int_{\gamma}\mathrm{d}\phi$$

for any 0-form ϕ defined on some differentiable curve $\gamma \subset \mathbb{R}^n$ (here the integral of ϕ εϖερ the boundary of the curve γ is understood to be the difference in value of ϕ βετωεεν the endpoints of the curve).

Notice the striking similarity between this statement and the generalized version of Stokes' theorem, which says that the integral of any compactly supported differential form ω over the boundary of some orientable manifold Ω is equal to the integral of its exterior derivative $d\omega$ over the whole of Ω, i.e.

$$\int_{\partial\Omega} \omega = \int_{\Omega} d\omega$$

This powerful statement is a generalization of the gradient theorem from 1-forms defined on one-dimensional manifolds to differential forms defined on manifolds of arbitrary dimension.

Green's Theorem

In mathematics, Green's theorem gives the relationship between a line integral around a simple closed curve C and a double integral over the plane region D bounded by C. It is the two-dimensional special case of the more general Stokes' theorem, and is named after British mathematician George Green.

Theorem

Let C be a positively oriented, piecewise smooth, simple closed curve in the plane ^{2}, and let D be the region bounded by C. If L and M are functions of (x, y) defined on an open region containing D and have continuous partial derivatives there, then

$$\oint_C (L dx + M dy) = \iint_D \left(\frac{\partial M}{\partial x} - \frac{\partial L}{\partial y} \right) dx dy.$$

For positive orientation, an arrow pointing in the counterclockwise direction may be drawn in the small circle in the integral symbol.

In physics, Green's theorem is mostly used to solve two-dimensional flow integrals, stating that the sum of fluid outflows at any point inside a volume is equal to the total outflow summed about an enclosing area. In plane geometry, and in particular, area surveying, Green's theorem can be used to determine the area and centroid of plane figures solely by integrating over the perimeter.

Proof When D is a Simple Region

If D is a simple region with its boundary consisting of the curves C_1, C_2, C_3, C_4, Green's theorem can be demonstrated.

The following is a proof of the theorem for the simplified area D, a type I region where C_2 and C_4 are vertical lines. A similar proof exists for when D is a type II region where C_1 and C_3 are straight lines. The general case can be deduced from this special case by approximating the domain D by a union of simple domains.

If it can be shown that

$$\oint_C L dx = \iint_D \left(-\frac{\partial L}{\partial y} \right) dA \qquad (1)$$

and

$$\oint_C M dy = \iint_D \left(\frac{\partial M}{\partial x} \right) dA \qquad (2)$$

are true, then Green's theorem is proven in the first case.

Define the type I region D as pictured on the right by

$$D = \{(x, y) \mid a \le x \le b, g_1(x) \le y \le g_2(x)\}$$

where g_1 and g_2 are continuous functions on $[a, b]$. Compute the double integral in (1):

$$\iint_D \frac{\partial L}{\partial y} dA = \int_a^b \int_{g_1(x)}^{g_2(x)} \frac{\partial L}{\partial y}(x, y) dy dx$$

$$= \int_a^b \{L(x, g_2(x)) - L(x, g_1(x))\} dx \qquad (3)$$

Now compute the line integral in (1). C can be rewritten as the union of four curves: C_1, C_2, C_3, C_4.

With C_1, use the parametric equations: $x = x$, $y = g_1(x)$, $a \le x \le b$. Then

$$\int_{C_1} L(x, y) dx = \int_a^b L(x, g_1(x)) dx$$

With C_3, use the parametric equations: $x = x$, $y = g_2(x)$, $a \le x \le b$. Then

$$\int_{C_3} L(x, y) dx = -\int_{-C_3} L(x, y) dx = \int_a^b L(x, g_2(x)) dx.$$

The integral over C_3 is negated because it goes in the negative direction from b to a, as C is oriented positively (counterclockwise). On C_2 and C_4, x remains constant, meaning

$$\int_{C_4} L(x, y) dx = \int_{C_2} L(x, y) dx = 0$$

Therefore,

$$\begin{aligned}\int_C L dx &= \int_{C_1} L(x,y)dx + \int_{C_2} L(x,y)dx + \int_{C_3} L(x,y)dx + \int_{C_4} L(x,y)dx \\ &= -\int_a^b L(x,g_2(x))dx + \int_a^b L(x,g_1(x))dx \qquad (4)\end{aligned}$$

Combining (3) with (4), we get (1). Similar computations give (2).

Relationship to the Stokes Theorem

Green's theorem is a special case of the Kelvin–Stokes theorem, when applied to a region in the xy-plane:

We can augment the two-dimensional field into a three-dimensional field with a z component that is always 0. Write F for the vector-valued function $F = (L, M, 0)$. Start with the left side of Green's theorem:

$$\oint_C (L dx + M dy) = \oint_C (L, M, 0)\cdot(dx, dy, dz) = \oint_C F\cdot dr.$$

Then by Kelvin–Stokes Theorem:

$$\oint_C F\cdot dr = \iint_S \nabla \times F\cdot\hat{n} dS.$$

The surface S is just the region in the plane D, with the unit normals $\hat{n}$ pointing up (in the positive z direction) to match the "positive orientation" definitions for both theorems.

The expression inside the integral becomes

$$\nabla \times F\cdot\hat{n} = \left[\left(\frac{\partial 0}{\partial y} - \frac{\partial M}{\partial z}\right)i + \left(\frac{\partial L}{\partial z} - \frac{\partial 0}{\partial x}\right)j + \left(\frac{\partial M}{\partial x} - \frac{\partial L}{\partial y}\right)k\right]\cdot k = \left(\frac{\partial M}{\partial x} - \frac{\partial L}{\partial y}\right).$$

Thus we get the right side of Green's theorem

$$\iint_S \nabla \times F\cdot\hat{n} dS = \iint_D \left(\frac{\partial M}{\partial x} - \frac{\partial L}{\partial y}\right) dA.$$

Relationship to the Divergence Theorem

Considering only two-dimensional vector fields, Green's theorem is equivalent to the following two-dimensional version of the divergence theorem:

$$\iint_D (\nabla\cdot F) dA = \oint_C F\cdot\hat{n} ds,$$

where $\hat{n}$ is the outward-pointing unit normal vector on the boundary.

To see this, consider the unit normal $\hat{n}$ in the right side of the equation. Since in Green's theorem $dr = (dx, dy)$ is a vector pointing tangential along the curve, and the curve C is the positively-oriented (i.e. counterclockwise) curve along the boundary, an outward normal would be a vector which points 90° to the right, which would be $(dy, -dx)$.

The length of this vector is $\sqrt{dx^2 + dy^2} = ds$.

So $\quad \hat{n}ds = (dy, -dx).$

Now let the components of $F = (P, Q)$. Then the right hand side becomes

$$\oint_C F \cdot \hat{n} ds = \oint_C P dy - Q dx$$

which by Green's theorem becomes

$$\oint_C -Q dx + P dy = \iint_D \left(\frac{\partial P}{\partial x} + \frac{\partial Q}{\partial y} \right) dA = \iint_D (\nabla \cdot F) dA.$$

The converse can also easily shown to be true.

Area Calculation

Green's theorem can be used to compute area by line integral. The area of D is given by:

$$A = \iint_D \mathrm{d}A.$$

Provided we choose L and M such that:

$$\frac{\partial M}{\partial x} - \frac{\partial L}{\partial y} = 1.$$

Then the area is given by:

$$A = \oint_C (L \mathrm{d}x + M \mathrm{d}y).$$

Possible formulas for the area of D include:

$$A = \oint_C x \mathrm{d}y = -\oint_C y \mathrm{d}x = \tfrac{1}{2} \oint_C (-y \mathrm{d}x + x \mathrm{d}y).$$

Chapter 8

Scalar Potential

A scalar potential is a fundamental concept in vector analysis and physics (the adjective *scalar* is frequently omitted if there is no danger of confusion with vector potential). The scalar potential is an example of a scalar field. Given a vector field F, the scalar potential *P* is defined such that:

$$F = -\nabla P = -\left(\frac{\partial P}{\partial x}, \frac{\partial P}{\partial y}, \frac{\partial P}{\partial z}\right),$$

where ∇ P is the gradient of P and the second part of the equation is minus the gradient for a function of the Cartesian coordinates x,y,z. In some cases, mathematicians may use a positive sign in front of the gradient to define the potential. Because of this definition of P in terms of the gradient, the direction of F at any point is the direction of the steepest decrease of P at that point, its magnitude is the rate of that decrease per unit length. In order for F to be described in terms of a scalar potential only, the following have to be true:

1. $-\int_a^b F \cdot dl = P(b) - P(a)$, where the integration is over a Jordan arc passing from location a to location b and P(b) is P evaluated at location b .
2. $\oint F \cdot dl = 0$, where the integral is over any simple closed path, otherwise known as a Jordan curve.
3. $\nabla \times F = 0$

The first of these conditions represents the fundamental theorem of the gradient and is true for any vector field that is a gradient of a differentiable single valued scalar field P. The second condition is a requirement of F so that it can be expressed as the gradient of a scalar function. The third condition re-expresses the second condition in terms of the curl of F using the fundamental theorem of the curl. A vector field F that satisfies these conditions is said to be irrotational (Conservative).

Scalar potentials play a prominent role in many areas of physics and engineering. The gravity potential is the scalar potential associated with the gravity per unit mass, i.e., the acceleration due to the field, as a function of position. The gravity potential is the gravitational potential energy per unit mass. In electrostatics the electric potential is the scalar potential associated with the electric field, i.e., with the electrostatic force per unit charge. The electric potential is in this case the electrostatic potential energy per unit charge. In fluid dynamics, irrotational lamellar fields have a scalar potential only in the special case when it is a Laplacian field. Certain aspects of the nuclear force can be described by a Yukawa potential. The potential play a prominent role in the Lagrangian and Hamiltonian formulations of classical mechanics. Further, the scalar potential is the fundamental quantity in quantum mechanics.

Not every vector field has a scalar potential. Those that do are called conservative, corresponding to the notion of conservative force in physics. Examples of non-conservative forces include frictional forces, magnetic forces, and in fluid mechanics a solenoidal field velocity field. By the Helmholtz decomposition theorem however, all vector fields can be describable in terms of a scalar potential and corresponding vector potential. In electrodynamics the electromagnetic scalar and vector potentials are known together as the electromagnetic four-potential.

Integrability Conditions

If F is a conservative vector field (also called *irrotational*, *curl-free*, or *potential*), and its components have continuous partial derivatives, the potential of F with respect to a reference point r_0 is defined in terms of the line integral:

$$V(r) = -\int_C F(r)\cdot dr = -\int_a^b F(r(t))\cdot r'(t)dt.$$

where C is a parametrized path from to

$$r(t), a \le t \le b, r(a) = r_0, r(b) = r.$$

The fact that the line integral depends on the path C only through its terminal points and is, in essence, the path independence property of a conservative vector field.

The fundamental theorem of calculus for line integrals implies that if V is defined in this way, then $F = -\nabla V$, so that V is a scalar potential of the conservative vector field F.

Scalar potential is not determined by the vector field alone: indeed, the gradient of a function is unaffected if a constant is added to it. If V is defined in terms of the line integral, the ambiguity of V reflects the freedom in the choice of the reference point r_0.

Altitude as Gravitational Potential Energy

An example is the (nearly) uniform gravitational field near the Earth's surface. It has a potential energy

$$U = mgh$$

where U is the gravitational potential energy and h is the height above the surface. This means that gravitational potential energy on a contour map is proportional to altitude. On a contour map, the two-dimensional negative gradient of the altitude is a two-dimensional vector field, whose vectors are always perpendicular to the contours and also perpendicular to the direction of gravity.

But on the hilly region represented by the contour map, the three-dimensional negative gradient of U always points straight downwards in the direction of gravity; F.

However, a ball rolling down a hill cannot move directly downwards due to the normal force of the hill's surface, which cancels out the component of gravity perpendicular to the hill's surface. The component of gravity that remains to move the ball is parallel to the surface:

$$F_S = -mg \sin\theta$$

where θ is the angle of inclination, and the component of F_S perpendicular to gravity is

$$F_P = -mg \sin\theta \cos\theta = -\frac{1}{2} mg \sin 2\theta.$$

This force F_p, parallel to the ground, is greatest when θ is 45 degrees. Let Δh be the uniform interval of altitude between contours on the contour map, and let Δx be the distance between two contours. Then

$$\theta = \tan^{-1}\frac{\Delta h}{\Delta x}$$

so that

$$F_p = -mg\frac{\Delta x \Delta h}{\Delta x^2 + \Delta h^2}.$$

However, on a contour map, the gradient is inversely proportional to Δx, which is not similar to force F_p: altitude on a contour map is not exactly a two-dimensional potential field.

The magnitudes of forces are different, but the directions of the forces are the same on a contour map as well as on the hilly region of the Earth's surface represented by the contour map.

Pressure as Buoyant Potential

In fluid mechanics, a fluid in equilibrium, but in the presence of a uniform gravitational field is permeated by a uniform buoyant force that cancels out the gravitational force: that is how the fluid maintains its equilibrium. This buoyant force is the negative gradient of pressure:

$$f_B = -\nabla p.$$

Since buoyant force points upwards, in the direction opposite to gravity, then pressure in the fluid increases downwards. Pressure in a static body of water increases proportionally to the depth below the surface of the water.

The surfaces of constant pressure are planes parallel to the ground. The surface of the water can be characterized as a plane with zero pressure. If the liquid has a vertical vortex (whose axis of rotation is perpendicular to the ground), then the vortex causes a depression in the pressure field. The surfaces of constant pressure are parallel to the ground far away from the vortex, but near and inside the vortex the surfaces of constant pressure are pulled downwards, closer to the ground. This also happens to the surface of zero pressure. Therefore, inside the vortex, the top surface of the liquid is pulled downwards into a depression, or even into a tube (a solenoid). The buoyant force due to a fluid on a solid object immersed and surrounded by that fluid can be obtained by integrating the negative pressure gradient along the surface of the object:

$$F_B = -\oint_S \nabla p \cdot dS.$$

A moving airplane wing makes the air pressure above it decrease relative to the air pressure below it. This creates enough buoyant force to counteract gravity.

Calculating the Scalar Potential

Given a vector field E, its scalar potential Φ can be calculated to be

$$\phi(R_0) = \frac{1}{4\pi}\int_\tau \frac{\nabla \cdot E(\tau)}{||R(\tau) - R_0||} d\tau$$

where τ is volume. Then, if E is irrotational (Conservative),

$$E = -\nabla\phi = -\frac{1}{4\pi}\nabla\int_\tau \frac{\nabla \cdot E(\tau)}{||R(\tau) - R_0||} d\tau.$$

This formula is known to be correct if E is continuous and vanishes asymptotically to zero towards infinity, decaying faster than $1/r$ and if the divergence of E likewise vanishes towards infinity, decaying faster than $1/r^2$.

Skew Gradient

In mathematics, a skew gradient of a harmonic function over a simply connected domain with two real dimensions is a vector field that is everywhere orthogonal to the gradient of the function, and has the same magnitude that the gradient has.

Definition

The skew gradient can be defined using complex analysis and the Cauchy–Riemann equations.

Let $f(z(x,y)) = u(x,y) + iv(x,y)$ be a complex-valued analytic function, where u, v are real-valued scalar functions of the real variables x, y.

A skew gradient is defined as:

$$\nabla^\perp u(x,y) = \nabla v(x,y)$$

and from the Cauchy–Riemann equations, it is derived that

$$\nabla^{\perp}u(x,y)=(-\frac{\partial u}{\partial y},\frac{\partial u}{\partial x})$$

Properties

The skew gradient has two interesting properties. It is everywhere orthogonal to the gradient of u, and of the same length:

$$\nabla u(x,y)\cdot\nabla^{\perp}u(x,y)=0, ||\nabla u||=||\nabla^{\perp}u||$$

Solenoidal Vector Field

In vector calculus a solenoidal vector field (also known as an incompressible vector field) is a vector field v with divergence zero at all points in the field:

$$\nabla\cdot v=0.$$

Properties

The fundamental theorem of vector calculus states that any vector field can be expressed as the sum of an irrotational and a solenoidal field. The condition of zero divergence is satisfied whenever a vector field v has only a vector potential component, because the definition of the vector potential A as:

$$v=\nabla\times A$$

automatically results in the identity (as can be shown, for example, using Cartesian coordinates):

$$\nabla\cdot v=\nabla\cdot(\nabla\times A)=0.$$

The converse also holds: for any solenoidal v there exists a vector potential A such that $v=\nabla\times A$. (Strictly speaking, this holds only subject to certain technical conditions on v.) The divergence theorem, gives the equivalent integral definition of a solenoidal field; namely that for any closed surface, the net total flux through the surface must be zero:

$$\oiint \mathrm{v}\cdot ds=0,$$

where ds is the outward normal to each surface element.

Etymology

Solenoidal has its origin in the Greek word for solenoid, which is meaning pipe-shaped, from pipe. In the present context of solenoidal it means constrained as if in a pipe, so with a fixed volume.

Examples

- the magnetic flux density B is solenoidal;
- the velocity field of an incompressible fluid flow is solenoidal;
- the vorticity field is solenoidal
- the electric flux density D in regions where there is no charge ($\rho_e = 0$);
- the current density J where the charge density is unvarying, $\frac{\partial \rho_e}{\partial t} = 0$.

Surface Gradient

In vector calculus, the surface gradient is a vector differential operator that is similar to the conventional gradient. The distinction is that the surface gradient takes effect along a surface. For a surface *S* in a scalar field u, the surface gradient is defined and notated as

$$\nabla_S u = \nabla u - \hat{n}(\hat{n} \cdot \nabla u)$$

where $\hat{n}$ is a unit normal to the surface. Examining the definition shows that the surface gradient is the (conventional) gradient with the component normal to the surface removed (subtracted), hence this gradient is tangent to the surface. In other words, the surface gradient is the orthographic projection of the gradient onto the surface.

The surface gradient arises whenever the gradient of a quantity over a surface is important. In the study of capillary surfaces for example, the gradient of spatially varying surface tension doesn't make much sense, however the surface gradient does and serves certain purposes.

Vector Field

In vector calculus, a vector field is an assignment of a vector to each point in a subset of Euclidean space. A vector field in the plane for instance can be visualized as an arrow, with a given magnitude and direction, attached to each point in the plane. Vector fields are often used to model, for example, the speed and direction of a moving fluid throughout space, or the strength and direction of some force, such as the magnetic or gravitational force, as it changes from point to point. The elements of differential and integral calculus extend to vector fields in a natural way. When a vector field represents force, the line integral of a vector field represents the work done by a force moving along a path, and under this interpretation conservation of energy is exhibited as a special case of the fundamental theorem of calculus.

Vector fields can usefully be thought of as representing the velocity of a moving flow in space, and this physical intuition leads to notions such as the divergence (which represents the rate of change of volume of a flow) and curl (which represents the rotation of a flow).

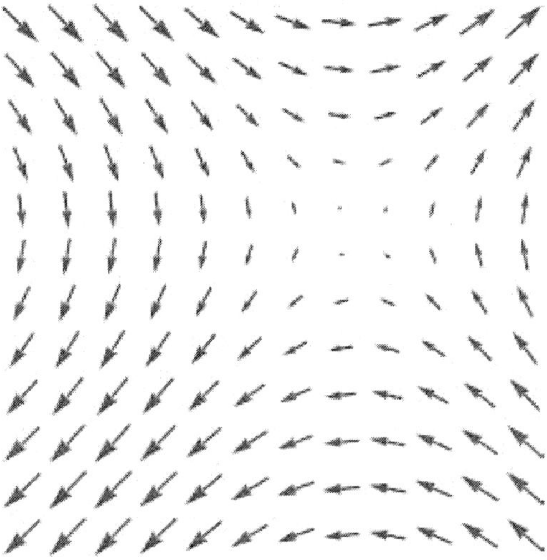

Figure: *A portion of the vector field (sin y, sin x)*

In coordinates, a vector field on a domain in n-dimensional Euclidean space can be represented as a vector-valued function that associates an n-tuple of real numbers to each point of the domain. This representation of a vector field depends on the coordinate system, and there is a well-defined transformation law in passing from one coordinate system to the other.

Vector fields are often discussed on open subsets of Euclidean space, but also make sense on other subsets such as surfaces, where they associate an arrow tangent to the surface at each point (a tangent vector).

More generally, vector fields are defined on differentiable manifolds, which are spaces that look like Euclidean space on small scales, but may have more complicated structure on larger scales. In this setting, a vector field gives a tangent vector at each point of the manifold (that is, a section of the tangent bundle to the manifold). Vector fields are one kind of tensor field.

Definition

Vector fields on subsets of Euclidean space:

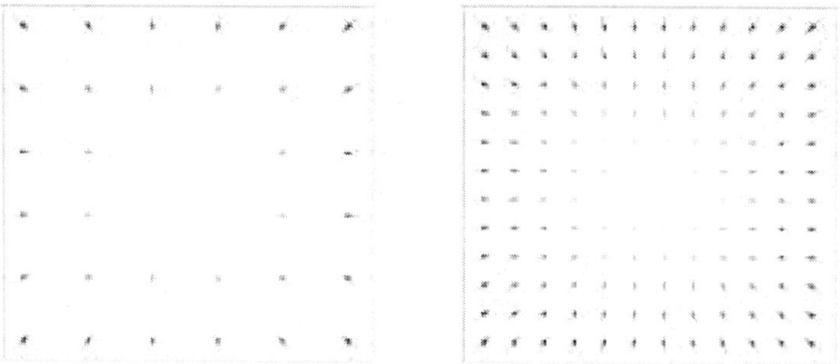

Figure: *Two representations of the same vector field:* $v(x, y) = -r$ *. The arrows depict the field at discrete points, however, the field exists everywhere.*

Given a subset S in R^n, a vector field is represented by a vector-valued function $V : S \rightarrow R^n$ in standard Cartesian coordinates (x_1, ..., x_n). If each component of V is continuous, then V is a continuous vector field, and more generally V is a C^k vector field if each component V is k times continuously differentiable.

A vector field can be visualized as assigning a vector to individual points within an n-dimensional space. Given two C^k-vector fields V, W defined on S and a real valued C^k-function f defined on S, the two operations scalar multiplication and vector addition

$$(fV)(p) := f(p)V(p)$$

$$(V + W)(p) := V(p) + W(p)$$

define the module of C^k-vector fields over the ring of C^k-functions.

Coordinate Transformation Law

In physics, a vector is additionally distinguished by how its coordinates change when one measures the same vector with respect to a different background coordinate system. The transformation properties of vectors distinguish a vector as a geometrically distinct entity from a simple list of scalars, or from a covector. Thus, suppose that $(x_1,...,x_n)$ is a choice of Cartesian coordinates, in terms of which the coordinates of the vector V are

$$V_x = (V_{1,x}, \ldots, V_{n,x})$$

and suppose that $(y_1,...,y_n)$ are n functions of the x_i defining a different coordinate system. Then the coordinates of the vector V in the new coordinates are required to satisfy the transformation law

$$V_{i,y} = \sum_{j=1}^{n} \frac{\partial y_j}{\partial x_i} V_{j,x}.$$

Such a transformation law is called contravariant. A similar transformation law characterizes vector fields in physics: specifically, a vector field is a specification of n functions in each coordinate system subject to the transformation law (1) relating the different coordinate systems.

Vector fields are thus contrasted with scalar fields, which associate a number or *scalar* to every point in space, and are also contrasted with simple lists of scalar fields, which do not transform under coordinate changes.

Vector Fields on Manifolds

Given a differentiable manifold M, a vector field on M is an assignment of a tangent vector to each point in M. More precisely, a vector field F is a mapping from M into the tangent bundle TM so that $p°F$ is the identity mapping where p denotes the projection from TM to M. In other words, a vector field is a section of the tangent bundle.

If the manifold M is smooth (respectively analytic)—that is, the change of coordinates are smooth (respectively analytic)—then one can make sense of the notion of smooth (respectively analytic) vector fields. The collection of all smooth vector fields on a smooth manifold M is often denoted by $\Gamma(\mathrm{T}M)$ or $C'(M,\mathrm{T}M)$ (especially when thinking of vector fields as sections); the collection of all smooth vector fields is also denoted by $\mathfrak{X}(M)$ (a fraktur "X").

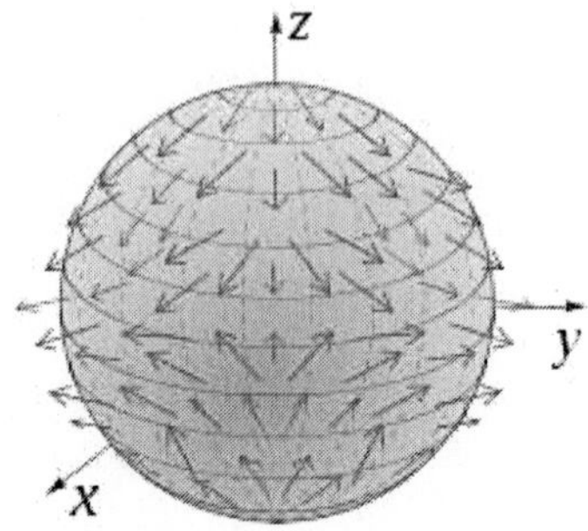

Figure: *A vector field on a sphere*

Examples

Figure: *The flow field around an airplane is a vector field in R^3, here visualized by bubbles that follow the streamlines showing a wingtip vortex.*

- A vector field for the movement of air on Earth will associate for every point on the surface of the Earth a vector with the wind speed and direction for that point. This can be drawn using arrows to represent the wind; the length (magnitude) of the arrow will be an indication of the wind speed. A "high" on the usual barometric pressure map would then act as a source (arrows pointing away), and a "low" would be a sink (arrows pointing towards), since air tends to move from high pressure areas to low pressure areas.

- Velocity field of a moving fluid. In this case, a velocity vector is associated to each point in the fluid.
- Streamlines, Streaklines and Pathlines are 3 types of lines that can be made from vector fields. They are :
  - streaklines — as revealed in wind tunnels using smoke.
  - streamlines (or fieldlines)— as a line depicting the instantaneous field at a given time.
  - pathlines — showing the path that a given particle (of zero mass) would follow.
- Magnetic fields. The fieldlines can be revealed using small iron filings.
- Maxwell's equations allow us to use a given set of initial conditions to deduce, for every point in Euclidean space, a magnitude and direction for the force experienced by a charged test particle at that point; the resulting vector field is the electromagnetic field.
- A gravitational field generated by any massive object is also a vector field. For example, the gravitational field vectors for a spherically symmetric body would all point towards the sphere's centre with the magnitude of the vectors reducing as radial distance from the body increases.

Gradient Field

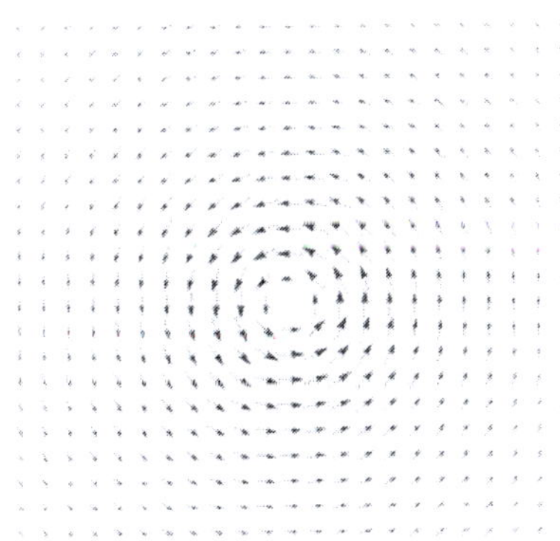

Figure: *A vector field that has circulation about a point cannot be written as the gradient of a function.*

Vector fields can be constructed out of scalar fields using the gradient operator (denoted by the del: ∇).

A vector field V defined on a set S is called a gradient field or a conservative field if there exists a real-valued function (a scalar field) f on S such that

$$V = \nabla f = \left(\frac{\partial f}{\partial x_1}, \frac{\partial f}{\partial x_2}, \frac{\partial f}{\partial x_3}, \ldots, \frac{\partial f}{\partial x_n}\right).$$

The associated flow is called the gradient flow, and is used in the method of gradient descent.

The path integral along any closed curve γ ($\gamma(0) = \gamma(1)$) in a gradient field is zero:

$$\int_\gamma \langle V(x), \mathrm{d}x\rangle = \int_\gamma \langle \nabla f(x), \mathrm{d}x\rangle = f(\gamma(1)) - f(\gamma(0)).$$

Central Field

A C^r-vector field over $\mathrm{R}^n \setminus \{0\}$ is called a central field if

$$V(T(p)) = T(V(p)) \qquad (T \in \mathrm{O}(n, R))$$

where $\mathrm{O}(n, \mathrm{R})$ is the orthogonal group. We say central fields are invariant under orthogonal transformations around 0.

The point 0 is called the centre of the field.

Since orthogonal transformations are actually rotations and reflections, the invariance conditions mean that vectors of a central field are always directed towards, or away from, 0; this is an alternate (and simpler) definition. A central field is always a gradient field, since defining it on one semiaxis and integrating gives an antigradient.

Operations on Vector Fields

Line Integral: A common technique in physics is to integrate a vector field along a curve, i.e. to determine its line integral. Given a particle in a gravitational vector field, where each vector represents the force acting on the particle at a given point in space, the line integral is the work done on the particle when it travels along a certain path.

The line integral is constructed analogously to the Riemann integral and it exists if the curve is rectifiable (has finite length) and the vector field is continuous.

Given a vector field V and a curve γ parametrized by [0, 1] the line integral is defined as

$$\int_\gamma \langle V(x), \mathrm{d}x\rangle = \int_0^1 \langle V(\gamma(t)), \gamma'(t)\, \mathrm{d}t\rangle.$$

Divergence

The divergence of a vector field on Euclidean space is a function (or scalar field). In three-dimensions, the divergence is defined by

$$\operatorname{div} F = \nabla\cdot F = \frac{\partial F_1}{\partial x} + \frac{\partial F_2}{\partial y} + \frac{\partial F_3}{\partial z},$$

with the obvious generalization to arbitrary dimensions. The divergence at a point represents the degree to which a small volume around the point is a source or a sink for the vector flow, a result which is made precise by the divergence theorem.

The divergence can also be defined on a Riemannian manifold, that is, a manifold with a Riemannian metric that measures the length of vectors.

Curl

The curl is an operation which takes a vector field and produces another vector field. The curl is defined only in three-dimensions, but some properties of the curl can be captured in higher dimensions with the exterior derivative. In three-dimensions, it is defined by

$$\operatorname{curl} F = \nabla\times F = \left(\frac{\partial F_3}{\partial y} - \frac{\partial F_2}{\partial z}\right)e_1 - \left(\frac{\partial F_3}{\partial x} - \frac{\partial F_1}{\partial z}\right)e_2 + \left(\frac{\partial F_2}{\partial x} - \frac{\partial F_1}{\partial y}\right)e_3.$$

The curl measures the density of the angular momentum of the vector flow at a point, that is, the amount to which the flow circulates around a fixed axis. This intuitive description is made precise by Stokes' theorem.

History

Vector fields arose originally in classical field theory in 19th century physics, specifically in magnetism. They were formalized by Michael Faraday, in his concept of *lines of force,* who emphasized that the field *itself* should be an object of study, which it has become throughout physics in the form of field theory.

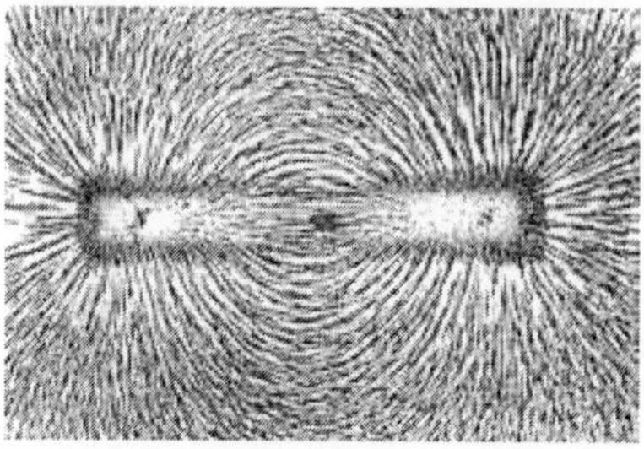

Figure: *Magnetic field lines of an iron bar (magnetic dipole)*

In addition to the magnetic field, other phenomena that were modelled as vector fields by Faraday include the electrical field and light field.

Flow Curves

Consider the flow of a fluid through a region of space. At any given time, any point of the fluid has a particular velocity associated with it; thus there is a vector field associated to any flow. The converse is also true: it is possible to associate a flow to a vector field having that vector field as its velocity.

Given a vector field V defined on S, one defines curves $\gamma(t)$ on S such that for each t in an interval I

$$\gamma'(t) = V(\gamma(t)).$$

By the Picard–Lindelöf theorem, if V is Lipschitz continuous there is a *unique* C^1-curve γ_x for each point x in S so that

$$\gamma_x(0) = x$$

$$\gamma_x'(t) = V(\gamma_x(t)) \qquad (t \in (-\epsilon, +\epsilon) \subset R).$$

The curves γ_x are called flow curves of the vector field V and partition S into equivalence classes. It is not always possible to extend the interval (-ε, +ε) to the whole real number line. The flow may for example reach the edge of S in a finite time.

In two or three dimensions one can visualize the vector field as giving rise to a flow on S. If we drop a particle into this flow at a point p it will move along the curve γ_p in the flow depending on the initial point p. If p is a stationary point of V then the particle will remain at p. Typical applications are streamline in fluid, geodesic flow, and one-parameter subgroups and the exponential map in Lie groups.

Complete Vector Fields

A vector field is complete if its flow curves exist for all time. In particular, compactly supported vector fields on a manifold are complete.

If X is a complete vector field on M, then the one-parameter group of diffeomorphisms generated by the flow along X exists for all time.

Difference Between Scalar and Vector Field

The difference between a scalar and vector field is not that a scalar is just one number while a vector is several numbers. The difference is in how their coordinates respond to coordinate transformations. A scalar *is* a coordinate whereas a vector *can be described* by coordinates, but it *is not* the collection of its coordinates.

Example

This example is about 2-dimensional Euclidean space (R^2) where we examine Euclidean (x, y) and polar (r, θ) coordinates (which are undefined at the origin). Thus $x = r \cos \theta$ and $y = r \sin \theta$ and also $r^2 = x^2 + y^2$, $\cos \theta = x/(x^2 + y^2)^{1/2}$ and $\sin \theta = y/(x^2 + y^2)^{1/2}$. Suppose we have a scalar field which is given by the constant function 1, and a vector field which attaches a vector in the r-direction with length 1 to each point. More precisely, they are given by the functions

$$s_{\text{polar}} : (r, \theta) \mapsto 1, \quad v_{\text{polar}} : (r, \theta) \mapsto (1, \theta).$$

Let us convert these fields to Euclidean coordinates. The vector of length 1 in the r-direction has the x coordinate $\cos \theta$ and the y coordinate $\sin \theta$. Thus in Euclidean coordinates the same fields are described by the functions

$$s_{\text{Euclidean}} : (x, y) \mapsto 1,$$

$$v_{\text{Euclidean}} : (x, y) \mapsto (\cos \theta, \sin \theta) = \left(\frac{x}{\sqrt{x^2 + y^2}}, \frac{y}{\sqrt{x^2 + y^2}} \right).$$

We see that while the scalar field remains the same, the vector field now looks different. The same holds even in the 1-dimensional case, as illustrated by the next example.

Example

Consider the 1-dimensional Euclidean space R with its standard Euclidean coordinate x. Suppose we have a scalar field and a vector field which are both given in the x coordinate by the constant function 1,

$$s_{\text{Euclidean}} : x \mapsto 1, \quad v_{\text{Euclidean}} : x \mapsto 1.$$

Thus, we have a scalar field which has the value 1 everywhere and a vector field which attaches a vector in the x-direction with magnitude 1 unit of x to each point.

Now consider the coordinate $\partial := 2x$. If x changes one unit then ∂ changes 2 units. Thus this vector field has a magnitude of 2 in units of ∂. Therefore, in the ∂ coordinate the scalar field and the vector field are described by the functions

$$s_{\text{unusual}} : \xi \mapsto 1, \quad v_{\text{unusual}} : \xi \mapsto 2$$

which are different.

Generalizations

Replacing vectors by p-vectors (pth exterior power of vectors) yields p-vector fields; taking the dual space and exterior powers yields differential k-forms, and combining these yields general tensor fields.

Algebraically, vector fields can be characterized as derivations of the algebra of smooth functions on the manifold, which leads to defining a vector field on a commutative algebra as a derivation on the algebra, which is developed in the theory of differential calculus over commutative algebras.

Chapter 9

Vector-valued Function

A vector-valued function also referred to as a vector function is a mathematical function of one or more variables whose range is a set of multidimensional vectors or infinite-dimensional vectors. The input of a vector-valued function could be a scalar or a vector. The dimension of the domain is not defined by the dimension of the range.

Properties

The domain of a vector-valued function is the intersection of the domain of the functions f, g, and h.

Derivative of a three-dimensional Vector Function

Many vector-valued functions, like scalar-valued functions, can be differentiated by simply differentiating the components in the Cartesian coordinate system. Thus, if

$$r(t) = f(t)i + g(t)j + h(t)k$$

is a vector-valued function, then

$$\frac{dr(t)}{dt} = f'(t)i + g'(t)j + h'(t)k.$$

The vector derivative admits the following physical interpretation: if r(t) represents the position of a particle, then the derivative is the velocity of the particle

$$v(t) = \frac{dr(t)}{dt}.$$

Likewise, the derivative of the velocity is the acceleration

$$\frac{dv(t)}{dt} = a(t).$$

Partial Derivative

The partial derivative of a vector function a with respect to a scalar variable q is defined as

$$\frac{\partial a}{\partial q} = \sum_{i=1}^{3} \frac{\partial a_i}{\partial q} e_i$$

where a_i is the *scalar component* of a in the direction of e_i. It is also called the direction cosine of a and e_i or their dot product. The vectors e_1, e_2, e_3 form an orthonormal basis fixed in the reference frame in which the derivative is being taken.

Ordinary Derivative

If a is regarded as a vector function of a single scalar variable, such as time t, then the equation above reduces to the first ordinary time derivative of a with respect to t,

$$\frac{da}{dt} = \sum_{i=1}^{3} \frac{da_i}{dt} e_i.$$

Total Derivative

If the vector a is a function of a number n of scalar variables q_r $(r = 1,...,n)$, and each q_r is only a function of time t, then the ordinary derivative of a with respect to t can be expressed, in a form known as the total derivative, as

$$\frac{da}{dt} = \sum_{r=1}^{n} \frac{\partial a}{\partial q_r}\frac{dq_r}{dt} + \frac{\partial a}{\partial t}.$$

Some authors prefer to use capital D to indicate the total derivative operator, as in *D/Dt*. The total derivative differs from the partial time derivative in that the total derivative accounts for changes in a due to the time variance of the variables q_r.

Reference Frames

Whereas for scalar-valued functions there is only a single possible reference frame, to take the derivative of a vector-valued function requires the choice of a reference frame (at least when a fixed Cartesian coordinate system is not implied as such). Once a reference frame has been chosen, the derivative of a vector-valued function can be computed using techniques similar to those for computing derivatives of scalar-valued functions. A different choice of reference frame will, in general, produce a different derivative function. The derivative functions in different reference frames have a specific kinematical relationship.

Derivative of a Vector Function with Nonfixed Bases

The above formulas for the derivative of a vector function rely on the assumption that the basis vectors e_1, e_2, e_3 are constant, that is, fixed in the reference frame in which the derivative of a is being taken, and therefore the e_1, e_2, e_3 each has a derivative of identically zero. This often holds true for problems dealing with vector fields in a fixed coordinate system, or for simple problems in physics. However, many complex problems involve the derivative of a vector function in multiple moving reference frames, which means that the basis vectors will not necessarily be constant. In such a case where the basis vectors e_1, e_2, e_3 are fixed in reference frame E, but not in reference frame N, the more general formula for the ordinary time derivative of a vector in reference frame N is

$$\frac{{}^{N}da}{dt} = \sum_{i=1}^{3} \frac{da_i}{dt} e_i + \sum_{i=1}^{3} a_i \frac{{}^{N}de_i}{dt}$$

where the superscript N to the left of the derivative operator indicates the reference frame in which the derivative is taken. As shown previously, the first term on the right hand side is equal to the derivative of a in the reference frame where e_1, e_2, e_3 are constant, reference frame E. It also can be shown that the second term on the right hand side is equal to the relative angular velocity of the two reference frames cross multiplied with the vector a itself. Thus, after substitution, the formula relating the derivative of a vector function in two reference frames is

$$\frac{{}^{N}da}{dt} = \frac{{}^{E}da}{dt} + {}^{N}\omega^{E} \times a$$

where ${}^{N}\hat{\omega}^{E}$ is the angular velocity of the reference frame E relative to the reference frame N.

One common example where this formula is used is to find the velocity of a space-borne object, such as a rocket, in the inertial reference frame using measurements of the rocket's velocity relative to the ground. The velocity $^{N}v^{R}$ in inertial reference frame N of a rocket R located at position r^{R} can be found using the formula

$$\frac{^{N}d}{dt}(r^{R}) = \frac{^{E}d}{dt}(r^{R}) + {}^{N}\omega^{E} \times r^{R}.$$

where $^{N}\dot{u}^{E}$ is the angular velocity of the Earth relative to the inertial frame N. Since velocity is the derivative of position, $^{N}v^{R}$ and $^{E}v^{R}$ are the derivatives of r^{R} in reference frames N and E, respectively. By substitution,

$$^{N}v^{R} = {}^{E}v^{R} + {}^{N}\omega^{E} \times r^{R}$$

where $^{E}v^{R}$ is the velocity vector of the rocket as measured from a reference frame E that is fixed to the Earth.

Derivative and Vector Multiplication

The derivative of the products of vector functions behaves similarly to the derivative of the products of scalar functions. Specifically, in the case of scalar multiplication of a vector, if p is a scalar variable function of q,

$$\frac{\partial}{\partial q}(pa) = \frac{\partial p}{\partial q}a + p\frac{\partial a}{\partial q}.$$

In the case of dot multiplication, for two vectors a and b that are both functions of q,

$$\frac{\partial}{\partial q}(a \cdot b) = \frac{\partial a}{\partial q} \cdot b + a \cdot \frac{\partial b}{\partial q}.$$

Similarly, the derivative of the cross product of two vector functions is

$$\frac{\partial}{\partial q}(a \times b) = \frac{\partial a}{\partial q} \times b + a \times \frac{\partial b}{\partial q}.$$

Derivative of an N-dimensional Vector Function

A function f of a real number t with values in the space R^{n} can be written as $f(t) = (f_1(t), f_2(t), \ldots, f_n(t))$. Its derivative equals

$$f'(t) = (f_{1'}(t), f_{2'}(t), \ldots, f_{n'}(t)).$$

If f is a function of several variables, say of $t \in R^m$, then the partial derivatives of the components of f form a $n \times m$ matrix called the *Jacobian matrix of f*.

Infinite-dimensional Vector Functions

If the values of a function f lie in an infinite-dimensional vector space X, such as a Hilbert space, then f may be called an *infinite-dimensional vector function.*

Functions with Values in a Hilbert Space

If the argument of f is a real number and X is a Hilbert space, then the derivative of f at a point t can be defined as in the finite-dimensional case:

$$f'(t) = \lim_{h \to 0} \frac{f(t+h) - f(t)}{h}.$$

Most results of the finite-dimensional case also hold in the infinite-dimensional case too, mutatis mutandis. Differentiation can also be defined to functions of several variables (e.g., $t \in R^n$ or even $t \in Y$, where Y is an infinite-dimensional vector space).

N.B. If X is a Hilbert space, then one can easily show that any derivative (and any other limit) can be computed componentwise: if

$$f = (f_1, f_2, f_3, \ldots)$$

(i.e., $f = f_1 e_1 + f_2 e_2 + f_3 e_3 + \cdots$, where $e_1, e_2, e_3, \ldots$ is an orthonormal basis of the space X), and $f'(t)$ exists, then

$$f'(t) = (f_{1'}(t), f_{2'}(t), f_{3'}(t), \ldots).$$

However, the existence of a componentwise derivative does not guarantee the existence of a derivative, as componentwise convergence in a Hilbert space does not guarantee convergence with respect to the actual topology of the Hilbert space.

Other infinite-dimensional vector spaces

Most of the above hold for other topological vector spaces X too. However, not as many classical results hold in the Banach space setting, e.g., an absolutely continuous function with values in a suitable Banach space need not have a derivative anywhere. Moreover, in most Banach spaces setting there are no orthonormal bases.

Integral Transform

In mathematics, an integral transform is any transform T of the following form:

$$(Tf)(u) = \int_{t_1}^{t_2} K(t,u) f(t) dt$$

The input of this transform is a function f, and the output is another function Tf. An integral transform is a particular kind of mathematical operator.

There are numerous useful integral transforms. Each is specified by a choice of the function K of two variables, the kernel function or nucleus of the transform.

Some kernels have an associated *inverse kernel* $K^{-1}(u, t)$ which (roughly speaking) yields an inverse transform:

$$f(t) = \int_{u_1}^{u_2} K^{-1}(u,t)(Tf(u))du$$

A *symmetric kernel* is one that is unchanged when the two variables are permuted.

Motivation

Mathematical notation aside, the motivation behind integral transforms is easy to understand. There are many classes of problems that are difficult to solve—or at least quite unwieldy algebraically—in their original representations. An integral transform "maps" an equation from its original "domain" into another domain. Manipulating and solving the equation in the target domain can be much easier than manipulation and solution in the original domain. The solution is then mapped back to the original domain with the inverse of the integral transform.

History

The precursor of the transforms were the Fourier series to express functions in finite intervals. Later the Fourier transform was developed to remove the requirement of finite intervals. Using the Fourier series, just about any practical function of time (the voltage across the terminals of an electronic device for example) can be represented as a sum of sines and cosines, each suitably scaled (multiplied by a constant factor), shifted (advanced or retarded in time) and "squeezed" or "stretched" (increasing or decreasing the frequency). The sines and cosines in the Fourier series are an example of an orthonormal basis.

Importance of Orthogonality

The individual basis functions have to be orthogonal. That is, the product of two dissimilar basis functions—integrated over their domain—must be zero. An integral transform, in actuality, just changes the representation of a function from one orthogonal basis to another. Each point in the representation of the transformed function in the image of the transform corresponds to the contribution of a given orthogonal basis function to the expansion. The process of expanding a function from its "standard" representation to a sum of a number of orthonormal basis functions, suitably scaled and shifted, is termed "spectral factorization." This is similar in concept to the description of a point in space in terms of three discrete components, namely, its x, y, and z coordinates. Each axis correlates only to itself and nothing to the other orthogonal axes. Note the terminological consistency: the determination of the amount by which an individual orthonormal basis function must be scaled in the spectral factorization of a function, F, is termed the "projection" of F onto that basis function.

The normal Cartesian graph *per se* of a function can be thought of as an orthonormal expansion. Indeed, each point just reflects the contribution of a given orthonormal basis function to the sum. In this way, the graph of a continuous real-valued function in the plane corresponds to an *infinite* set of basis functions; if the number of basis functions were finite, the curve would consist of a discrete set of points rather than a continuous contour.

Usage Example

As an example of an application of integral transforms, consider the Laplace transform. This is a technique that maps differential or integro-differential equations in the "time" domain into polynomial equations in what is termed the "complex frequency" domain. (Complex frequency is similar to actual, physical frequency but rather more general. Specifically, the imaginary component ω of the complex frequency $s = -\sigma + i\omega$ corresponds to the usual concept of frequency, *viz.*, the speed at which a sinusoid cycles, whereas the real component σ of the complex frequency corresponds to the degree of "damping".)

The equation cast in terms of complex frequency is readily solved in the complex frequency domain (roots of the polynomial equations in the complex frequency domain correspond to eigenvalues in the time domain), leading to a "solution" formulated in the frequency domain. Employing the inverse transform, *i.e.*, the inverse procedure of the original Laplace transform, one obtains a time-domain solution. In this example, polynomials in the complex frequency domain (typically occurring in the denominator) correspond to power series in the time domain, while axial shifts in the complex frequency domain correspond to damping by decaying exponentials in the time domain.

The Laplace transform finds wide application in physics and particularly in electrical engineering, where the characteristic equations that describe the behaviour of an electric circuit in the complex frequency domain correspond to linear combinations of exponentially damped, scaled, and time-shifted sinusoids in the time domain. Other integral transforms find special applicability within other scientific and mathematical disciplines.

Different Domains

Here integral transforms are defined for functions on the real numbers, but they can be defined more generally for functions on a group.

- If instead one uses functions on the circle (periodic functions), integration kernels are then biperiodic functions; convolution by functions on the circle yields circular convolution.
- If one uses functions on the cyclic group of order n (C_n or Z/nZ), one obtains $n \times n$ matrices as integration kernels; convolution corresponds to circulant matrices.

General Theory

Although the properties of integral transforms vary widely, they have some properties in common. For example, every integral transform is a linear operator, since the integral is a linear operator, and in fact if the kernel is allowed to be a generalized function then all linear operators are integral transforms (a properly formulated version of this statement is the Schwartz kernel theorem).

The general theory of such integral equations is known as Fredholm theory. In this theory, the kernel is understood to be a compact operator acting on a Banach space of functions. Depending on the situation, the kernel is then variously referred to as the Fredholm operator, the nuclear operator or the Fredholm kernel.

Order of Integration

Order of integration, denoted *I(d)*, is a summary statistic for a time series. It reports the minimum number of differences required to obtain a stationary series.

Integration of Order Zero

A time series is integrated of order 0 if it admits a moving average representation with

$$\sum_{k=0}^{\infty} |\gamma_k^2| < \infty$$

which means that the autocovariance is decaying to 0 sufficiently fast. This is a necessary, but not sufficient condition for a stationary process. Therefore, all stationary processes are I(0), but not all I(0) processes are stationary.

Integration of Order D

A time series is integrated of order d if

$$(1-L)^d X_t$$

is integrated of order 0, where L is the lag operator and $(1-L)$ is the first difference, ie: $(1-L)X_t = X_t - X_{t-1} = \Delta X$.

In other words, a process is integrated to order d if taking repeated differences d times yields a stationary process.

Constructing an Integrated Series

An *I*(*d*) process can be constructed by summing an *I*(*d*"1) process:

- Suppose X_t is $I(d\text{-}1)$
- Now construct a series $Z_t = \sum_{k=0}^{t} X_k$
- Show that Z is $I(d)$ by observing its first-differences are $I(d\text{-}1)$:

$$\Delta Z_t = \sum_{k=0}^{t} X_k - \sum_{k=0}^{t-1} X_k = X_t \sim I(d-1)$$

Coordinate System

In geometry, a coordinate system is a system which uses one or more numbers, or coordinates, to uniquely determine the position of a point or other geometric element on a manifold such as Euclidean space. The order of the coordinates is significant and they are sometimes identified by their position in an ordered tuple and sometimes by a letter, as in 'the *x*-coordinate'.

In elementary mathematics the coordinates are taken to be real numbers, but may be complex numbers or elements of a more abstract system such as a commutative ring. The use of a coordinate system allows problems in geometry to be translated into problems about numbers and *vice versa*; this is the basis of analytic geometry.

An example in everyday use is the system of assigning longitude and latitude to geographical locations. In physics, a coordinate system used to describe points in space is called a frame of reference.

Number Line

In basic mathematics, a number line is a picture of a straight line on which every point is assumed to correspond to a real number and every real number to a point. Often the integers are shown as specially-marked points evenly spaced on the line. Although this image only shows the integers from "9 to 9, the line includes all real numbers, continuing forever in each direction, and also numbers not marked that are between the integers. It is often used as an aid in teaching simple addition and subtraction, especially involving negative numbers.

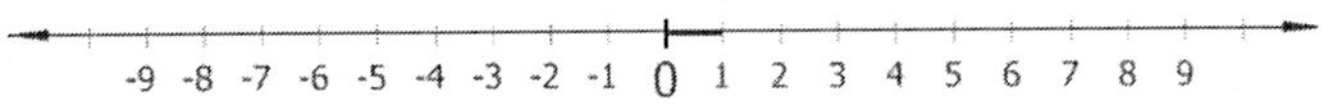

Figure: *It is divided into two symmetric halves by the origin, i.e. the number zero.*

In advanced mathematics, the expressions real number line, or real line are typically used to indicate the above-mentioned concept that every point on a straight line corresponds to a single real number, and vice versa.

Drawing the Number Line

The number line is usually represented as being horizontal. Customarily, positive numbers lie on the right side of zero, and negative numbers lie on the left side of zero. An arrowhead on either end of the drawing is meant to suggest that the line continues indefinitely in the positive and negative real numbers, denoted by $\mathbb{R}$. The real numbers consist of irrational numbers and rational numbers, as well as the integers, whole numbers, and the natural numbers (the counting numbers).

A line drawn through the origin at right angles to the real number line can be used to represent the imaginary numbers. This line, called imaginary line, extends the number line to a complex number plane, with points representing complex numbers.

Chapter 10

Quadrants and Octants

The axes of a two-dimensional Cartesian system divide the plane into four infinite regions, called quadrants, each bounded by two half-axes.

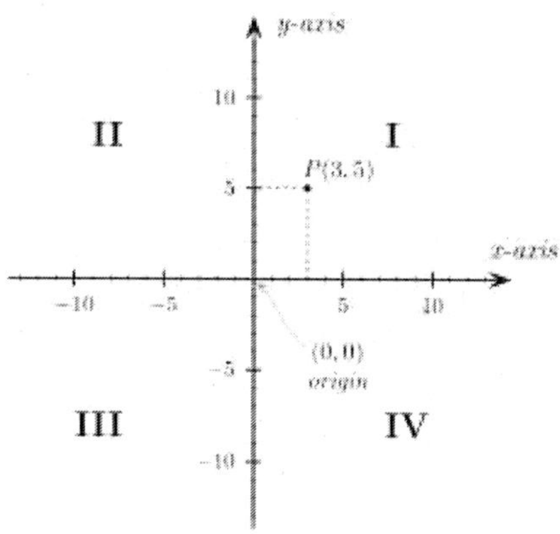

Figure: *The four quadrants of a Cartesian coordinate system.*

These are often numbered from 1st to 4th and denoted by Roman numerals: I (where the signs of the two coordinates are I (+,+), II(-,+), III (-,-), and IV (+,-). When the axes are drawn according to the mathematical custom, the numbering goes counter-clockwise starting from the upper right ("northeast") quadrant.

Similarly, a three-dimensional Cartesian system defines a division of space into eight regions or octants, according to the signs of the coordinates of the points. The octant where all three coordinates are positive is sometimes called the first octant; however, there is no established nomenclature for the other octants. The n-dimensional generalization of the quadrant and octant is the orthant.

Cartesian Space

A Euclidean plane with a chosen Cartesian system is called a Cartesian plane. Since Cartesian coordinates are unique and non-ambiguous, the points of a Cartesian plane can be identified with all possible pairs of real numbers; that is with the Cartesian product $\mathbb{R}^2 = \mathbb{R} \times \mathbb{R}$, where $\mathbb{R}$ is the set of all reals. In the same way one defines a Cartesian space of any dimension n, whose points can be identified with the tuples (lists) of n real numbers, that is, with $\mathbb{R}^n$.

Cartesian Formulas for the Plane

Distance Between two Points: The Euclidean distance between two points of the plane with Cartesian coordinates (x_1, y_1) and (x_2, y_2) is

$$d = \sqrt{(x_2 - x_1)^2 + (y_2 - y_1)^2}.$$

This is the Cartesian version of Pythagoras' theorem. In three-dimensional space, the distance between points (x_1, y_1, z_1) and (x_2, y_2, z_2) is

$$d = \sqrt{(x_2 - x_1)^2 + (y_2 - y_1)^2 + (z_2 - z_1)^2}$$

which can be obtained by two consecutive applications of Pythagoras' theorem.

Euclidean Transformations

The Euclidean transformations or Euclidean motions are the (bijective) mappings of points of the Euclidean plane to themselves which preserve distances between points. There are four types of these mappings (also called isometries): translations, rotations, reflections and glide reflections.

Translation

Translating a set of points of the plane, preserving the distances and directions between them, is equivalent to adding a fixed pair of numbers (a,b) to the Cartesian coordinates of every point in the set. That is, if the original coordinates of a point are (x,y), after the translation they will be

$$(x',y') = (x+a, y+b).$$

Rotation

To rotate a figure counterclockwise around the origin by some angle is equivalent to replacing every point with coordinates (x,y) by the point with coordinates (x',y'), where

$$x' = x\cos\theta - y\sin\theta$$

$$y' = x\sin\theta + y\cos\theta.$$

Thus: $(x',y') = ((x\cos\theta - y\sin\theta), (x\sin\theta + y\cos\theta))$.

Reflection

If (x, y) are the Cartesian coordinates of a point, then $(-x, y)$ are the coordinates of its reflection across the second coordinate axis (the Y axis), as if that line were a mirror. Likewise, $(x, -y)$ are the coordinates of its reflection across the first coordinate axis (the X axis). In more generality, reflection across a line through the origin making an angle with the x-axis, is equivalent to replacing every point with coordinates (x,y) by the point with coordinates (x',y'), where

$$x' = x\cos 2\theta + y\sin 2\theta$$

$$y' = x\sin 2\theta - y\cos 2\theta.$$

Thus: $(x',y') = ((x\cos 2\theta + y\sin 2\theta), (x\sin 2\theta - y\cos 2\theta))$.

Glide Reflection

A glide reflection is the composition of a reflection across a line followed by a translation in the direction of that line. It can be seen that the order of these operations does not matter (the translation can come first, followed by the reflection).

General Matrix form of the Transformations

These Euclidean transformations of the plane can all be described in a uniform way by using matrices. The result (x',y') of applying a Euclidean transformation to a point (x,y) is given by the formula

$$(x',y') = (x,y)A + b$$

where A is a 2×2 orthogonal matrix and $b = (b_1, b_2)$ is an arbitrary ordered pair of numbers; that is,

$$x' = xA_{11} + yA_{21} + b_1$$

$$y' = xA_{12} + yA_{22} + b_2,$$

where

$A = \begin{pmatrix} A_{11} & A_{12} \\ A_{21} & A_{22} \end{pmatrix}$. [Note the use of row vectors for point coordinates and that the matrix is written on the right.]

To be *orthogonal*, the matrix A must have orthogonal rows with same Euclidean length of one, that is,

$$A_{11}A_{21} + A_{12}A_{22} = 0$$

and $$A_{11}^2 + A_{12}^2 = A_{21}^2 + A_{22}^2 = 1.$$

This is equivalent to saying that A times its transpose must be the identity matrix. If these conditions do not hold, the formula describes a more general affine transformation of the plane provided that the determinant of A is not zero.

The formula defines a translation if and only if A is the identity matrix. The transformation is a rotation around some point if and only if A is a rotation matrix, meaning that

$$A_{11}A_{22} - A_{21}A_{12} = 1.$$

A reflection or glide reflection is obtained when,

$$A_{11}A_{22} - A_{21}A_{12} = -1.$$

Assuming that translation is not used transformations can be combined by simply multiplying the associated transformation matrices.

Affine Transformation

Another way to represent coordinate transformations in Cartesian coordinates is through affine transformations. In affine transformations an extra dimension is added and all points are given a value of 1 for this extra dimension.

The advantage of doing this is that then all of the euclidean transformations become linear transformations and can be represented using matrix multiplication. The affine transformation is given by:

$$\begin{pmatrix} A_{11} & A_{21} & b_1 \\ A_{12} & A_{22} & b_2 \\ 0 & 0 & 1 \end{pmatrix}\begin{pmatrix} x \\ y \\ 1 \end{pmatrix} = \begin{pmatrix} x' \\ y' \\ 1 \end{pmatrix}.$$

[Note the A matrix from above was transposed. The matrix is on the left and column vectors for point coordinates are used.] Using affine transformations multiple different euclidean transformations including translation can be combined by simply multiplying the corresponding matrices.

Scaling

An example of an affine transformation which is not a Euclidean motion is given by scaling.

To make a figure larger or smaller is equivalent to multiplying the Cartesian coordinates of every point by the same positive number m.

If (x,y) are the coordinates of a point on the original figure, the corresponding point on the scaled figure has coordinates

$$(x', y') = (mx, my).$$

If m is greater than 1, the figure becomes larger; if m is between 0 and 1, it becomes smaller.

Shearing

A shearing transformation will push the top of a square sideways to form a parallelogram. Horizontal shearing is defined by:

$$(x', y') = (x + ys, y)$$

Shearing can also be applied vertically:

$$(x', y') = (x, xs + y)$$

In Two Dimensions

Fixing or choosing the x-axis determines the y-axis up to direction. Namely, the y-axis is necessarily the perpendicular to the x-axis through the point marked 0 on the x-axis. But there is a choice of which of the two half lines on the perpendicular to designate as positive and which as negative.

Each of these two choices determines a different orientation (also called *handedness*) of the Cartesian plane.

The usual way of orienting the axes, with the positive x-axis pointing right and the positive y-axis pointing up (and the x-axis being the "first" and the y-axis the "second" axis) is considered the *positive* or *standard* orientation, also called the *right-handed* orientation.

A commonly used mnemonic for defining the positive orientation is the *right hand rule.*

Placing a somewhat closed right hand on the plane with the thumb pointing up, the fingers point from the x-axis to the y-axis, in a positively oriented coordinate system.

The other way of orienting the axes is following the *left hand rule*, placing the left hand on the plane with the thumb pointing up.

When pointing the thumb away from the origin along an axis, the curvature of the fingers indicates a positive rotation along that axis.

Regardless of the rule used to orient the axes, rotating the coordinate system will preserve the orientation. Switching any two axes will reverse the orientation.

In Three Dimensions

Once the x- and y-axes are specified, they determine the line along which the z-axis should lie, but there are two possible directions on this line.

The two possible coordinate systems which result are called 'right-handed' and 'left-handed'.

The standard orientation, where the xy-plane is horizontal and the z-axis points up (and the x- and the y-axis form a positively oriented two-dimensional coordinate system in the xy-plane if observed from *above* the xy-plane) is called right-handed or positive.

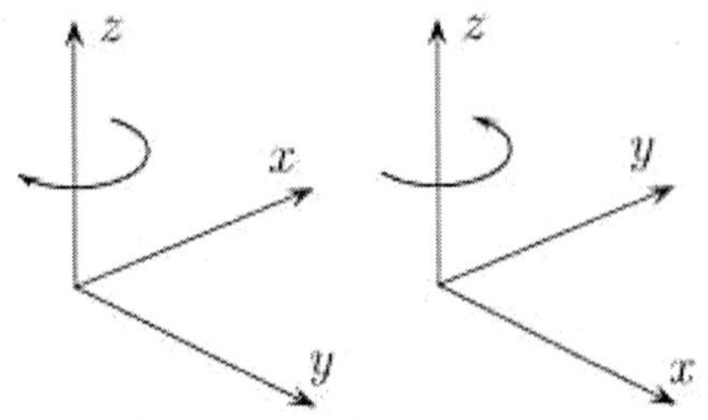

***Figure:** The left-handed orientation is shown on the left, and the right-handed on the right.*

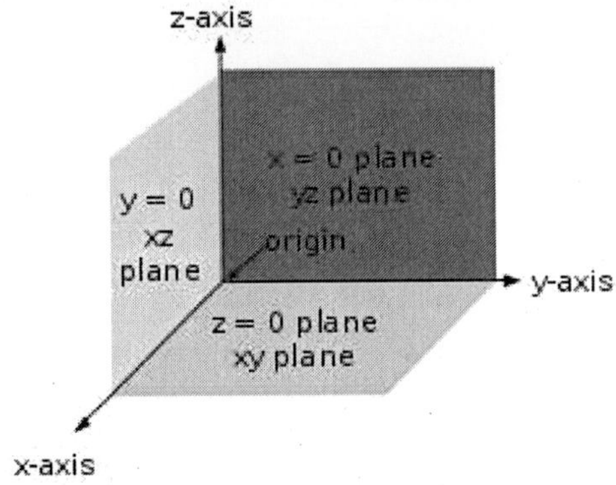

***Figure:** The right-handed Cartesian coordinate system indicating the coordinate planes.*

The name derives from the right-hand rule. If the index finger of the right hand is pointed forward, the middle finger bent inward at a right angle to it, and the thumb placed at a right angle to both, the three fingers indicate the relative directions of the x-, y-, and z-axes in a *right-handed* system. The thumb indicates the x-axis, the index finger the y-axis and the middle finger the z-axis. Conversely, if the same is done with the left hand, a left-handed system results.

Figure depicted earlier, depicts a left and a right-handed coordinate system. Because a three-dimensional object is represented on the two-dimensional screen, distortion and ambiguity result. The axis pointing downward (and to the right) is also meant to point *towards* the observer, whereas the "middle" axis is meant to point *away* from the observer. The red circle is *parallel* to the horizontal xy-plane and indicates rotation from the x-axis to the y-axis (in both cases). Hence the red arrow passes *in front of* the z-axis. Figure presented is another attempt at depicting a right-handed coordinate system. Again, there is an ambiguity caused by projecting the three-dimensional coordinate system into the plane. Many observers see Figure as "flipping in and out" between a convex cube and a concave "corner". This corresponds to the two possible orientations of the coordinate system. Seeing the figure as convex gives a left-handed coordinate system. Thus the "correct" way to view Figure given is to imagine the x-axis as pointing *towards* the observer and thus seeing a concave corner.

Representing a Vector in the Standard Basis

A point in space in a Cartesian coordinate system may also be represented by a position vector, which can be thought of as an arrow pointing from the origin of the coordinate system to the point. If the coordinates represent spatial positions (displacements), it is common to represent the vector from the origin to the point of interest as . In two dimensions, the vector from the origin to the point with Cartesian coordinates (x, y) can be written as:

$$r = xi + yj$$

where $i = (1,0)$, and $j = (0,1)$ are unit vectors in the direction of the x-axis and y-axis respectively, generally referred to as the *standard basis* (in some application areas these may also be referred to as versors). Similarly, in three dimensions, the vector from the origin to the point with Cartesian coordinates (x, y, z) can be written as:

$$r = xi + yj + zk$$

where $k = (0,0,1)$ is the unit vector in the direction of the z-axis.

There is no *natural* interpretation of multiplying vectors to obtain another vector that works in all dimensions, however there is a way to use complex numbers to provide such a multiplication. In a two dimensional cartesian plane, identify the point with coordinates (x, y) with the complex number $z = x + iy$. Here, i is the complex number whose square is the real number "1 and is identified with the point with coordinates (0,1), so it is not the unit vector in the direction of the x-axis (this confusion is just an unfortunate historical accident). Since the complex numbers can be multiplied giving another complex number, this identification provides a means to "multiply" vectors. In a three dimensional cartesian space a similar identification can be made with a subset of the quaternions.

Applications

Each axis may have different units of measurement associated with it (such as kilograms, seconds, pounds, etc.). Although four- and higher-dimensional spaces are difficult to visualize, the algebra of Cartesian coordinates can be extended relatively easily to four or more variables, so that certain calculations involving many variables can be done. (This sort of algebraic extension is what is used to define the geometry of higher-dimensional spaces.) Conversely, it is often helpful to use the geometry of Cartesian coordinates in two or three dimensions to visualize algebraic relationships between two or three of many non-spatial variables.

The graph of a function or relation is the set of all points satisfying that function or relation. For a function of one variable, f, the set of all points (x,y) where $y = f(x)$ is the graph of the function f. For a function of two variables, g, the set of all points (x,y,z) where $z = g(x,y)$ is the graph of the function g. A sketch of the graph of such a function or relation would consist of all the salient parts of the function or relation which would include its relative extrema, its concavity and points of inflection, any points of discontinuity and its end behaviour. All of these terms are more fully defined in calculus. Such graphs are useful in calculus to understand the nature and behaviour of a function or relation.

Note that positions on a surface in navigation use latitude and longitude in a similar two dimensional system. However the co-ordinates are written in the opposite sequence, effectively (y,x).

Polar Coordinate System

In mathematics, the polar coordinate system is a two-dimensional coordinate system in which each point on a plane is determined by a distance from a fixed point and an angle from a fixed direction.

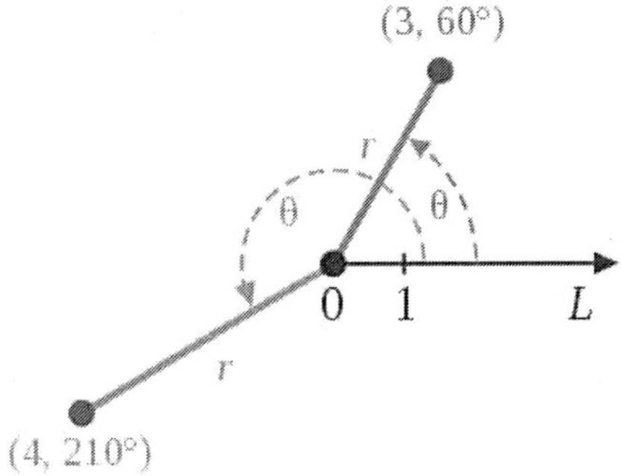

Figure: *Points in the polar coordinate system with pole O and polar axis L. In green, the point with radial coordinate 3 and angular coordinate 60 degrees, or (3,60°). In blue, the point (4,210°).*

The fixed point (analogous to the origin of a Cartesian system) is called the *pole*, and the ray from the pole in the fixed direction is the *polar axis*. The distance from the pole is called the *radial coordinate* or *radius*, and the angle is the *angular coordinate*, *polar angle*, or *azimuth*.

History

The concepts of angle and radius were already used by ancient peoples of the 1st millennium BCE. The Greek astronomer and astrologer Hipparchus (190–120 BCE) created a table of chord functions giving the length of the chord for each angle, and there are references to his using polar coordinates in establishing stellar positions. In *On Spirals*, Archimedes describes the Archimedean spiral, a function whose radius depends on the angle. The Greek work, however, did not extend to a full coordinate system.

From the 8th century CE onward, astronomers developed methods for approximating and calculating the direction to Makkah (qibla)—and its distance—from any location on the Earth. From the 9th century onward they were using spherical trigonometry and map projection methods to determine these quantities accurately.

The calculation is essentially the conversion of the equatorial polar coordinates of Mecca (i.e. its longitude and latitude) to its polar coordinates (i.e. its qibla and distance) relative to a system whose reference meridian is the great circle through the given location and the Earth's poles, and whose polar axis is the line through the location and its antipodal point. There are various accounts of the introduction of polar coordinates as part of a formal coordinate system. The full history of the subject is described in Harvard professor Julian Lowell Coolidge's *Origin of Polar Coordinates.* Grégoire de Saint-Vincent and Bonaventura Cavalieri independently introduced the concepts in the mid-seventeenth century.

Saint-Vincent wrote about them privately in 1625 and published his work in 1647, while Cavalieri published his in 1635 with a corrected version appearing in 1653. Cavalieri first used polar coordinates to solve a problem relating to the area within an Archimedean spiral. Blaise Pascal subsequently used polar coordinates to calculate the length of parabolic arcs.

In *Method of Fluxions* (written 1671, published 1736), Sir Isaac Newton examined the transformations between polar coordinates, which he referred to as the "Seventh Manner; For Spirals", and nine other coordinate systems.

In the journal *Acta Eruditorum* (1691), Jacob Bernoulli used a system with a point on a line, called the *pole* and *polar axis* respectively. Coordinates were specified by the distance from the pole and the angle from the *polar axis*. Bernoulli's work extended to finding the radius of curvature of curves expressed in these coordinates. The actual term *polar coordinates* has been attributed to Gregorio Fontana and was used by 18th-century Italian writers. The term appeared in English in George Peacock's 1816 translation of Lacroix's *Differential and Integral Calculus*. Alexis Clairaut was the first to think of polar coordinates in three dimensions, and Leonhard Euler was the first to actually develop them.

Conventions

The radial coordinate is often denoted by r, and the angular coordinate by θ or t.

Angles in polar notation are generally expressed in either degrees or radians (2π rad being equal to 360°). Degrees are traditionally used in navigation, surveying, and many applied disciplines, while radians are more common in mathematics and mathematical physics.

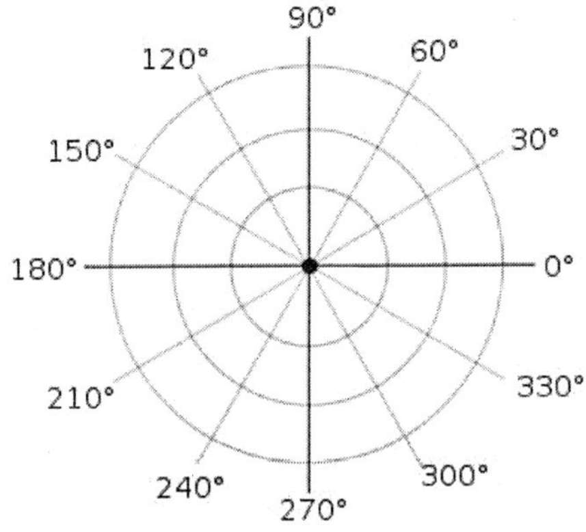

***Figure:** A polar grid with several angles labelled in degrees*

In many contexts, a positive angular coordinate means that the angle θ is measured counterclockwise from the axis.

In mathematical literature, the polar axis is often drawn horizontal and pointing to the right.

Uniqueness of Polar Coordinates

Adding any number of full turns (360°) to the angular coordinate does not change the corresponding direction. Also, a negative radial coordinate is best interpreted as the corresponding positive distance measured in the opposite direction. Therefore, the same point can be expressed with an infinite number of different polar coordinates $(r, \theta \pm n \times 360°)$ or $(-r, \theta \pm (2n + 1)180°)$, where n is any integer. Moreover, the pole itself can be expressed as $(0, \theta)$ for any angle θ.

Where a unique representation is needed for any point, it is usual to limit r to non-negative numbers ($r \geq 0$) and θ to the interval [0, 360°) or (-180°, 180°] (in radians, $[0, 2\pi)$ or $(-\pi, \pi]$). One must also choose a unique azimuth for the pole, e.g., $\theta = 0$.

Converting Between Polar and Cartesian Coordinates

The two polar coordinates r and θ can be converted to the two Cartesian coordinates x and y by using the trigonometric functions sine and cosine:

$$x = r\cos\theta$$

$$y = r\sin\theta$$

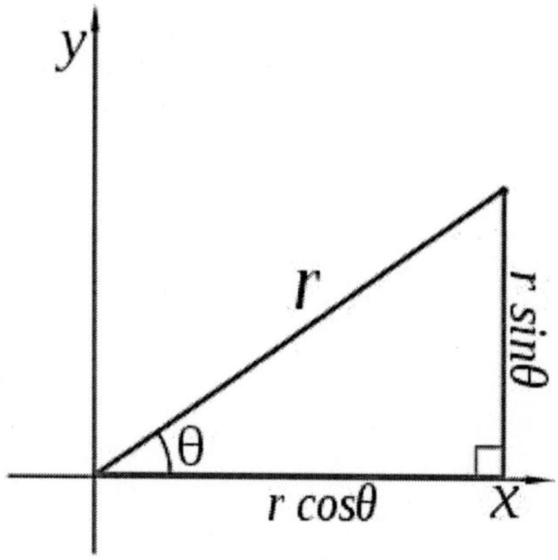

Figure: *A diagram illustrating the relationship between polar and Cartesian coordinates.*

The Cartesian coordinates x and y can be converted to polar coordinates r and θ with $r \geq 0$ and θ in the interval $(-\pi, \pi]$ by:

$$r = \sqrt{y^2 + x^2} \quad \text{(as in the Pythagorean theorem), and}$$

$$\theta = \operatorname{atan} 2(y, x)$$ (where atan2 is a common variation on the arctangent function that takes into account the quadrant) or

$$\theta = \begin{cases} \arctan(|\frac{y}{x}|) & \text{if } x > 0 \\ \arctan(|\frac{y}{x}|) + \pi & \text{if } x < 0 \text{ and } y \geq 0 \\ \arctan(|\frac{y}{x}|) - \pi & \text{if } x < 0 \text{ and } y < 0 \\ \frac{\pi}{2} & \text{if } x = 0 \text{ and } y > 0 \\ -\frac{\pi}{2} & \text{if } x = 0 \text{ and } y < 0 \\ 0 & \text{if } x = 0 \text{ and } y = 0 \end{cases}$$

In degrees this would be from -180° to 180°. If desired an angle in the range $[0, 2\pi)$ may be obtained by adding 2π to the value if it is negative. The zero angle at the origin where x and y are both zero is just a convenient value that is often chosen.

The value of θ above is the principal value of the complex number function arg applied to $x+iy$, except that arg does not define a value at the origin.

Polar Equation of a Curve

The equation defining an algebraic curve expressed in polar coordinates is known as a *polar equation*. In many cases, such an equation can simply be specified by defining r as a function of θ. The resulting curve then consists of points of the form $(r(\theta), \theta)$ and can be regarded as the graph of the polar function r.

Different forms of symmetry can be deduced from the equation of a polar function r. If $r(-\theta) = r(\theta)$ the curve will be symmetrical about the horizontal (0°/180°) ray, if $r(\pi - \theta) = r(\theta)$ it will be symmetric about the vertical (90°/270°) ray, and if $r(\theta - \alpha) = r(\theta)$ it will be rotationally symmetric α counterclockwise about the pole.

Because of the circular nature of the polar coordinate system, many curves can be described by a rather simple polar equation, whereas their Cartesian form is much more intricate. Among the best known of these curves are the polar rose, Archimedean spiral, lemniscate, limaçon, and cardioid. For the circle, line, and polar rose below, it is understood that there are no restrictions on the domain and range of the curve.

Circle

The general equation for a circle with a centre at (r_0, φ) and radius a is

$$r^2 - 2rr_0\cos(\theta - \varphi) + r_0^2 = a^2.$$

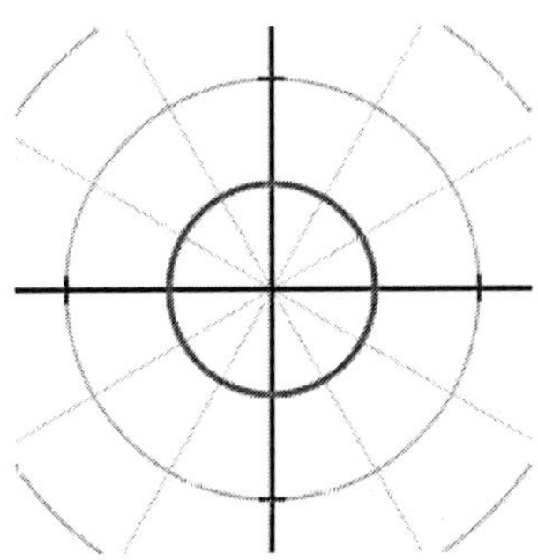

Figure: *A circle with equation* $r(\theta) = 1$

This can be simplified in various ways, to conform to more specific cases, such as the equation

$$r(\theta) = a$$

for a circle with a centre at the pole and radius *a*.

When r_0 = a, or when the origin lies on the circle, the equation becomes

$$r = 2a\cos(\theta - \varphi).$$

In the general case, the equation can be solved for *r*, giving

$$r = r_0 \cos(\theta - \varphi) + \sqrt{a^2 - r_0^2 \sin^2(\theta - \varphi)},$$

the solution with a minus sign in front of the square root giving the same curve.

Line

Radial lines (those running through the pole) are represented by the equation

$$\theta = \varphi,$$

where φ is the angle of elevation of the line; that is, φ = arctan *m* where *m* is the slope of the line in the Cartesian coordinate system. The non-radial line that crosses the radial line $\theta = \varphi$ perpendicularly at the point (r_0, φ) has the equation

$$r(\theta) = r_0 \sec(\theta - \varphi).$$

Otherwise stated (r_0, φ) is the point in which the tangent intersects the imaginary circle of radius r_0.

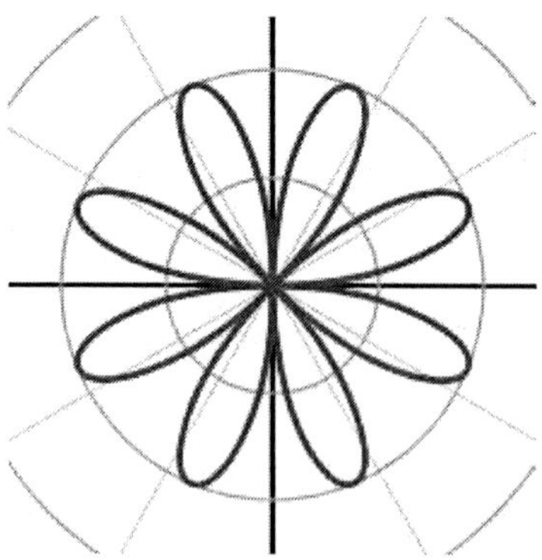

Figure: *A polar rose with equation $r(\theta) = 2 \sin 4\theta$*

Polar Rose

A polar rose is a famous mathematical curve that looks like a petaled flower, and that can be expressed as a simple polar equation,

$$r(\theta) = a\cos(k\theta + \varphi_0)$$

for any constant φ_0 (including 0). If k is an integer, these equations will produce a k-petaled rose if k is odd, or a $2k$-petaled rose if k is even. If k is rational but not an integer, a rose-like shape may form but with overlapping petals. Note that these equations never define a rose with 2, 6, 10, 14, etc. petals. The variable a represents the length of the petals of the rose.

Archimedean Spiral

The Archimedean spiral is a famous spiral that was discovered by Archimedes, which also can be expressed as a simple polar equation. It is represented by the equation

$$r(\theta) = a + b\theta.$$

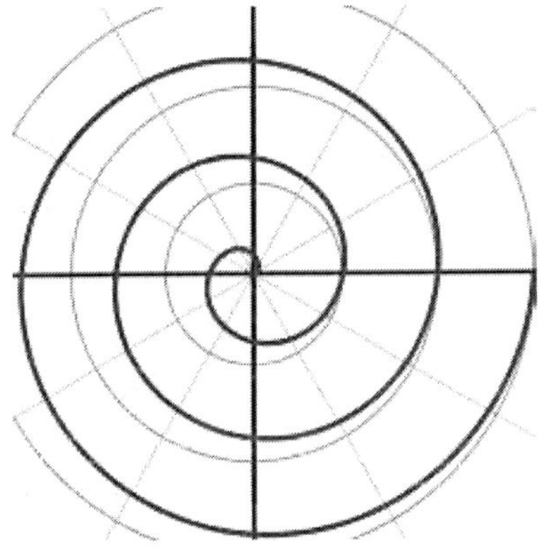

Figure: *One arm of an Archimedean spiral with equation $r(\theta) = \theta / 2\pi$ for $0 < \theta < 6\pi$*

Changing the parameter a will turn the spiral, while b controls the distance between the arms, which for a given spiral is always constant. The Archimedean spiral has two arms, one for $\theta > 0$ and one for $\theta < 0$. The two arms are smoothly connected at the pole. Taking the mirror image of one arm across the 90°/ 270° line will yield the other arm.

This curve is notable as one of the first curves, after the conic sections, to be described in a mathematical treatise, and as being a prime example of a curve that is best defined by a polar equation.

Conic Sections

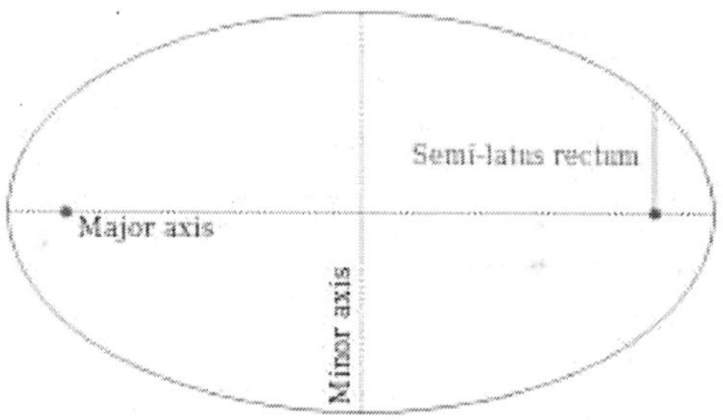

Figure: *Ellipse, showing semi-latus rectum*

A conic section with one focus on the pole and the other somewhere on the 0° ray (so that the conic's major axis lies along the polar axis) is given by:

$$r = \frac{\ell}{1 + e\cos\theta}$$

where e is the eccentricity and is the semi-latus rectum (the perpendicular distance at a focus from the major axis to the curve). If $e > 1$, this equation defines a hyperbola; if $e = 1$, it defines a parabola; and if $e < 1$, it defines an ellipse. The special case $e = 0$ of the latter results in a circle of radius ℓ.

Complex Numbers

Every complex number can be represented as a point in the complex plane, and can therefore be expressed by specifying either the point's Cartesian coordinates (called rectangular or Cartesian form) or the point's polar coordinates (called polar form). The complex number z can be represented in rectangular form as

$$z = x + iy$$

where i is the imaginary unit, or can alternatively be written in polar form (via the conversion formulae given above) as

$$z = r\cdot(\cos\theta + i\sin\theta)$$

and from there as

$$z = re^{i\theta}$$

where e is Euler's number, which are equivalent as shown by Euler's formula. (Note that this formula, like all those involving exponentials of angles, assumes that the angle θ is expressed in radians.) To convert between the rectangular and polar forms of a complex number, the conversion formulae given above can be used.

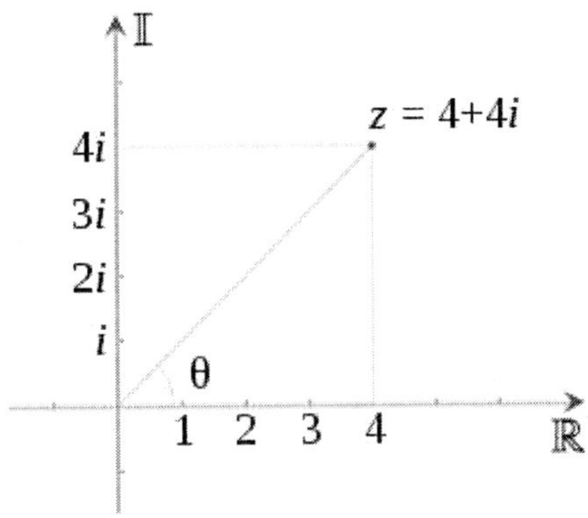

Figure: *An illustration of a complex number z plotted on the complex plane*

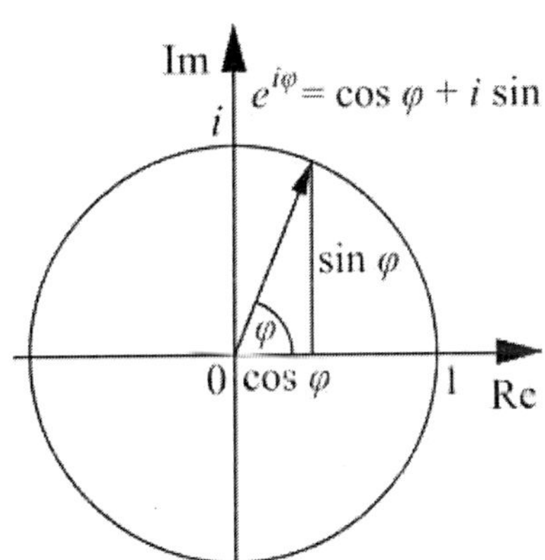

Figure: *An illustration of a complex number plotted on the complex*
plane using Euler's formula

For the operations of multiplication, division, and exponentiation of complex numbers, it is generally much simpler to work with complex numbers expressed in polar form rather than rectangular form. From the laws of exponentiation:

- Multiplication:

$$r_0 e^{i\theta_0} \cdot r_1 e^{i\theta_1} = r_0 r_1 e^{i(\theta_0+\theta_1)}$$

- Division:

$$\frac{r_0 e^{i\theta_0}}{r_1 e^{i\theta_1}} = \frac{r_0}{r_1} e^{i(\theta_0-\theta_1)}$$

- Exponentiation (De Moivre's formula):

$$(re^{i\theta})^n = r^n e^{in\theta}$$

Calculus

Calculus can be applied to equations expressed in polar coordinates.

The angular coordinate θ is expressed in radians throughout this section, which is the conventional choice when doing calculus.

Differential Calculus

Using $x = r \cos \theta$ and $y = r \sin \theta$, one can derive a relationship between derivatives in Cartesian and polar coordinates. For a given function, $u(x,y)$, it follows that

$$r\frac{\partial u}{\partial r} = r\frac{\partial u}{\partial x}\frac{\partial x}{\partial r} + r\frac{\partial u}{\partial y}\frac{\partial y}{\partial r},$$

$$\frac{\partial u}{\partial \theta} = \frac{\partial u}{\partial x}\frac{\partial x}{\partial \theta} + \frac{\partial u}{\partial y}\frac{\partial y}{\partial \theta},$$

or

$$r\frac{\partial u}{\partial r} = r\frac{\partial u}{\partial x}\cos\theta + r\frac{\partial u}{\partial y}\sin\theta = x\frac{\partial u}{\partial x} + y\frac{\partial u}{\partial y},$$

$$\frac{\partial u}{\partial \theta} = -\frac{\partial u}{\partial x} r\sin\theta + \frac{\partial u}{\partial y} r\cos\theta = -y\frac{\partial u}{\partial x} + x\frac{\partial u}{\partial y}.$$

Hence, we have the following formulae:

$$r\frac{\partial}{\partial r} = x\frac{\partial}{\partial x} + y\frac{\partial}{\partial y}$$

$$\frac{\partial}{\partial\theta} = -y\frac{\partial}{\partial x} + x\frac{\partial}{\partial y}.$$

To find the Cartesian slope of the tangent line to a polar curve $r(\theta)$ at any given point, the curve is first expressed as a system of parametric equations.

$$x = r(\theta)\cos\theta$$

$$y = r(\theta)\sin\theta$$

Differentiating both equations with respect to θ yields

$$\frac{dx}{d\theta} = r'(\theta)\cos\theta - r(\theta)\sin\theta$$

$$\frac{dy}{d\theta} = r'(\theta)\sin\theta + r(\theta)\cos\theta.$$

Dividing the second equation by the first yields the Cartesian slope of the tangent line to the curve at the point $(r(\theta), \theta)$:

$$\frac{dy}{dx} = \frac{r'(\theta)\sin\theta + r(\theta)\cos\theta}{r'(\theta)\cos\theta - r(\theta)\sin\theta}.$$

Integral Calculus (Arc Length)

The arc length (length of a line segment) defined by a polar function is found by the integration over the curve $r(\theta)$. Let L denote this length along the curve starting from points A through to point B, where these points correspond to $\theta = a$ and $\theta = b$ such that $0 < b$ “ $a < 2\pi$. The length of L is given by the following integral

$$L = \int_a^b \sqrt{\left[r(\theta)\right]^2 + \left[\frac{dr(\theta)}{d\theta}\right]^2}\, d\theta$$

Integral Calculus (Area)

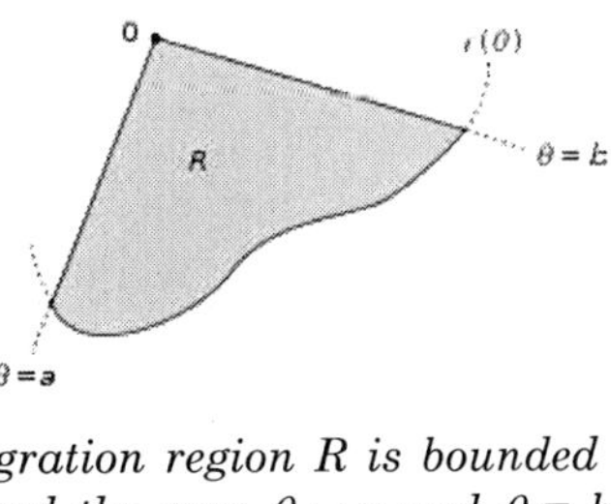

Figure: *The integration region R is bounded by the curve $r(\theta)$ and the rays $\theta = a$ and $\theta = b$.*

Let R denote the region enclosed by a curve $r(\theta)$ and the rays $\theta = a$ and $\theta = b$, where $0 < b - a < 2\pi$. Then, the area of R is

$$\frac{1}{2}\int_a^b [r(\theta)]^2 \, d\theta.$$

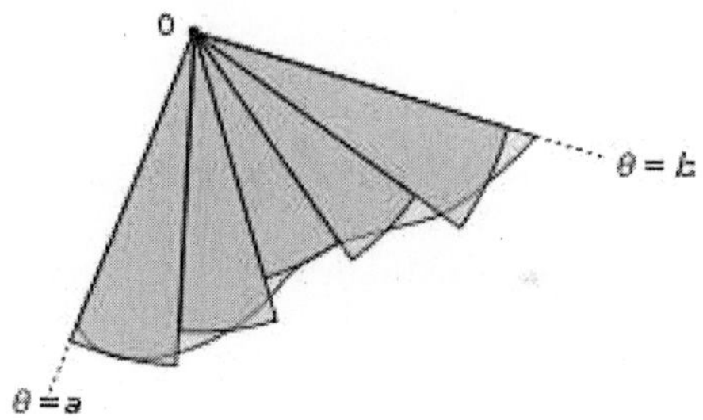

Figure: *The region R is approximated by n sectors (here, n = 5).*

This result can be found as follows. First, the interval $[a, b]$ is divided into n subintervals, where n is an arbitrary positive integer. Thus $\Delta\theta$, the length of each subinterval, is equal to $b - a$ (the total length of the interval), divided by n, the number of subintervals. For each subinterval $i = 1, 2, \ldots, n$, let θ_i be the midpoint of the subinterval, and construct a sector with the centre at the pole, radius $r(\theta_i)$, central angle $\Delta\theta$ and arc length $r(\theta_i)\Delta\theta$. The area of each constructed sector is therefore equal to

$$[r(\theta_i)]^2 \pi \cdot \frac{\Delta\theta}{2\pi} = \frac{1}{2}[r(\theta_i)]^2 \Delta\theta.$$

Figure: *A planimetre, which mechanically computes polar integrals*

Hence, the total area of all of the sectors is

$$\sum_{i=1}^{n} \tfrac{1}{2} r(\theta_i)^2 \, \Delta\theta.$$

As the number of subintervals n is increased, the approximation of the area continues to improve. In the limit as $n \to \infty$, the sum becomes the Riemann sum for the above integral.

A mechanical device that computes area integrals is the planimetre, which measures the area of plane figures by tracing them out: this replicates integration in polar coordinates by adding a joint so that the 2-element linkage effects Green's theorem, converting the quadratic polar integral to a linear integral.

Generalization

Using Cartesian coordinates, an infinitesimal area element can be calculated as $dA = dx\ dy$. The substitution rule for multiple integrals states that, when using other coordinates, the Jacobian determinant of the coordinate conversion formula has to be considered:

$$J = \det\frac{\partial(x,y)}{\partial(r,\theta)} = \begin{vmatrix} \frac{\partial x}{\partial r} & \frac{\partial x}{\partial \theta} \\ [8pt]\frac{\partial y}{\partial r} & \frac{\partial y}{\partial \theta} \end{vmatrix} = \begin{vmatrix} \cos\theta & -r\sin\theta \\ \sin\theta & r\cos\theta \end{vmatrix} = r\cos^2\theta + r\sin^2\theta = r.$$

Hence, an area element in polar coordinates can be written as

$$dA = dx\,dy = J\,dr\,d\theta = r\,dr\,d\theta.$$

Now, a function that is given in polar coordinates can be integrated as follows:

$$\iint_R f(r,0)\,dA = \int_a^b \int_0^{r(\theta)} f(r,\theta)\,r\,dr\,d\theta.$$

Here, R is the same region as above, namely, the region enclosed by a curve $r(\theta)$ and the rays $\theta = a$ and $\theta = b$.

The formula for the area of R mentioned above is retrieved by taking f identically equal to 1. A more surprising application of this result yields the Gaussian integral

$$\int_{-\infty}^{\infty} e^{-x^2}\,dx = \sqrt{\pi}.$$

Vector Calculus

Vector calculus can also be applied to polar coordinates. For a planar motion, let r be the position vector ($r\cos(\theta)$, $r\sin(\theta)$), with r and θ depending on time t.

We define the unit vectors

$$\hat{r} = (\cos(\theta), \sin(\theta))$$

in the direction of r and

$$\hat{\theta} = (-\sin(\theta), \cos(\theta)) = \hat{k} \times \hat{r},$$

in the plane of the motion perpendicular to the radial direction, where $\hat{k}$ is a unit vector normal to the plane of the motion.

Then

$$r = (x, y) = r(\cos\theta, \sin\theta) = r\hat{r},$$

$$\dot{r} = (\dot{x}, \dot{y}) = \dot{r}(\cos\theta, \sin\theta) + r\dot{\theta}(-\sin\theta, \cos\theta) = \dot{r}\hat{r} + r\dot{\theta}\hat{\theta},$$

$$\ddot{r} = (\ddot{x}, \ddot{y}) = \ddot{r}(\cos\theta, \sin\theta) + 2\dot{r}\dot{\theta}(-\sin\theta, \cos\theta) + r\ddot{\theta}(-\sin\theta, \cos\theta) - r\dot{\theta}^2(\cos\theta, \sin\theta) =$$

$$\left(\ddot{r} - r\dot{\theta}^2\right)\hat{r} + \left(2\dot{r}\dot{\theta} + r\ddot{\theta}\right)\hat{\theta} = (\ddot{r} - r\dot{\theta}^2)\hat{r} + \frac{1}{r}\frac{d}{dt}\left(r^2\dot{\theta}\right)\hat{\theta} = (\ddot{r} - r\dot{\theta}^2)\hat{r} + (r\ddot{\theta} + 2\dot{r}\dot{\theta})\hat{\theta}$$

Centrifugal and Coriolis Terms

The term $r\dot{\theta}^2$ is sometimes referred to as the *centrifugal term*, and the term $2\dot{r}\dot{\theta}$ as the *Coriolis term*.

Although these equations bear some resemblance in form to the centrifugal and Coriolis effects found in rotating reference frames, nonetheless these are not the same things. For example, the physical centrifugal and Coriolis forces appear only in non-inertial frames of reference.

In contrast, these terms that appear when acceleration is expressed in polar coordinates are a mathematical consequence of differentiation; these terms appear wherever polar coordinates are used.

In particular, these terms appear even when polar coordinates are used in inertial frames of reference, where the physical centrifugal and Coriolis forces never appear.

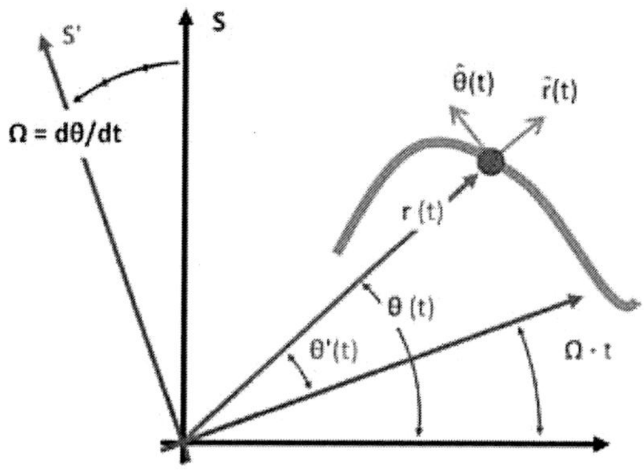

Figure: *Inertial frame of reference S and instantaneous non-inertial co-rotating frame of reference S2 . The co-rotating frame rotates at angular rate Ω equal to the rate of rotation of the particle about the origin of S2 at the particular moment t. Particle is located at vector position r(t) and unit vectors are shown in the radial direction to the particle from the origin, and also in the direction of increasing angle θ normal to the radial direction. These unit vectors need not be related to the tangent and normal to the path. Also, the radial distance r need not be related to the radius of curvature of the path.*

Co-rotating Frame

For a particle in planar motion, one approach to attaching physical significance to these terms is based on the concept of an instantaneous *co-rotating frame of reference*. To define a co-rotating frame, first an origin is selected from which the distance *r(t)* to the particle is defined. An axis of rotation is set up that is perpendicular to the plane of motion of the particle, and passing through this origin.

Then, at the selected moment *t*, the rate of rotation of the co-rotating frame Ω is made to match the rate of rotation of the particle about this axis, *dè/dt*. Next, the terms in the acceleration in the inertial frame are related to those in the co-rotating frame. Let the location of the particle in the inertial frame be *(r(*t*), θ(*t*)), and in the co-rotating frame be (*r(t), *θ2* (t)*). Because the co-rotating frame rotates at the same rate as the particle,* dθ2 /dt = *0. The fictitious centrifugal force in the co-rotating frame is* $mr\Omega^2$, radially outward. The velocity of the particle in the co-rotating frame also is radially outward, because $d\theta 2 / dt = 0$. The *fictitious Coriolis force* therefore has a value $-2m(dr/dt)\Omega$, pointed in the direction of increasing θ only. Thus, using these forces in Newton's second law we find:

$$F + F_{cf} + F_{Cor} = m\ddot{r},$$

where over dots represent time differentiations, and F is the net real force (as opposed to the fictitious forces). In terms of components, this vector equation becomes:

$$F_r + mr\Omega^2 = m\ddot{r}$$

$$F_\theta - 2m\dot{r}\Omega = mr\ddot{\theta},$$

which can be compared to the equations for the inertial frame:

$$F_r = m\ddot{r} - mr\dot{\theta}^2$$

$$F_\theta = mr\ddot{\theta} + 2m\dot{r}\dot{\theta}.$$

This comparison, plus the recognition that by the definition of the co-rotating frame at time t it has a rate of rotation $\Omega = d\theta/dt$, shows that we can interpret the terms in the acceleration (multiplied by the mass of the particle) as found in the inertial frame as the negative of the centrifugal and Coriolis forces that would be seen in the instantaneous, non-inertial co-rotating frame.

For general motion of a particle (as opposed to simple circular motion), the centrifugal and Coriolis forces in a particle's frame of reference commonly are referred to the instantaneous osculating circle of its motion, not to a fixed centre of polar coordinates.

Connection to Spherical and Cylindrical Coordinates

The polar coordinate system is extended into three dimensions with two different coordinate systems, the cylindrical and spherical coordinate system.

Applications

Polar coordinates are two-dimensional and thus they can be used only where point positions lie on a single two-dimensional plane. They are most appropriate in any context where the phenomenon being considered is inherently tied to direction and length from a centre point. For instance, the examples above show how elementary polar equations suffice to define curves—such as the Archimedean spiral—whose equation in the Cartesian coordinate system would be much more intricate. Moreover, many physical systems—such as those concerned with bodies moving around a central point or with phenomena originating from a central point—are simpler and more intuitive to model using polar coordinates. The initial motivation for the introduction of the polar system was the study of circular and orbital motion.

Position and Navigation

Polar coordinates are used often in navigation, as the destination or direction of travel can be given as an angle and distance from the object being considered. For instance, aircraft use a slightly modified version of the polar coordinates for navigation. In this system, the one generally used for any sort of navigation, the 0° ray is generally called heading 360, and the angles continue in a clockwise direction, rather than counterclockwise, as in the mathematical system. Heading 360 corresponds to magnetic north, while headings 90, 180, and 270 correspond to magnetic east, south, and west, respectively. Thus, an aircraft travelling 5 nautical miles due east will be travelling 5 units at heading 90 (read zero-niner-zero by air traffic control).

Modelling

Systems displaying radial symmetry provide natural settings for the polar coordinate system, with the central point acting as the pole. A prime example of this usage is the groundwater flow equation when applied to radially symmetric wells. Systems with a radial force are also good candidates for the use of the polar coordinate system. These systems include gravitational fields, which obey the inverse-square law, as well as systems with point sources, such as radio antennas.

Radially asymmetric systems may also be modelled with polar coordinates. For example, a microphone's pickup pattern illustrates its proportional response to an incoming sound from a given direction, and these patterns can be represented as polar curves. The curve for a standard cardioid microphone, the most common unidirectional microphone, can be represented as $r = 0.5 + 0.5\sin(\theta)$ at its target design frequency. The pattern shifts toward omnidirectionality at lower frequencies.

Cylindrical Coordinate System

A cylindrical coordinate system is a three-dimensional coordinate system that specifies point positions by the distance from a chosen reference axis, the direction from the axis relative to a chosen reference direction, and the distance from a chosen reference plane perpendicular to the axis. The latter distance is given as a positive or negative number depending on which side of the reference plane faces the point.

The *origin* of the system is the point where all three coordinates can be given as zero. This is the intersection between the reference plane and the axis.

The axis is variously called the *cylindrical* or *longitudinal* axis, to differentiate it from the *polar axis*, which is the ray that lies in the reference plane, starting at the origin and pointing in the reference direction.

The distance from the axis may be called the *radial distance* or *radius*, while the angular coordinate is sometimes referred to as the *angular position* or as the *azimuth*. The radius and the azimuth are together called the *polar coordinates*, as they correspond to a two-dimensional polar coordinate system in the plane through the point, parallel to the reference plane. The third coordinate may be called the *height* or *altitude* (if the reference plane is considered horizontal), *longitudinal position*, or *axial position*.

Cylindrical coordinates are useful in connection with objects and phenomena that have some rotational symmetry about the longitudinal axis, such as water flow in a straight pipe with round cross-section, heat distribution in a metal cylinder, electromagnetic fields produced by an electric current in a long, straight wire, and so on.

It is sometimes called "cylindrical polar coordinate" and "polar cylindrical coordinate", and is sometime used to specify the position of stars in a galaxy ("galactocentric cylindrical polar coordinate").

Definition

The three coordinates (ρ, φ, z) of a point P are defined as:

- The radial distance ρ is the Euclidean distance from the z axis to the point P.
- The azimuth φ is the angle between the reference direction on the chosen plane and the line from the origin to the projection of P on the plane.
- The height z is the signed distance from the chosen plane to the point P.

Unique Cylindrical Coordinates

As in polar coordinates, the same point with cylindrical coordinates (ρ, φ, z) has infinitely many equivalent coordinates, namely (ρ, $\varphi \pm n\times360°$, z) and ($-\rho$, $\varphi \pm (2n + 1)\times180°$, z), where n is any integer. Moreover, if the radius ρ is zero, the azimuth is arbitrary.

In situations where one needs a unique set of coordinates for each point, one may restrict the radius to be non-negative ($\rho \geq 0$) and the azimuth φ to lie in a specific interval spanning 360°, such as (-180°,+180°] or [0,360°).

Conventions

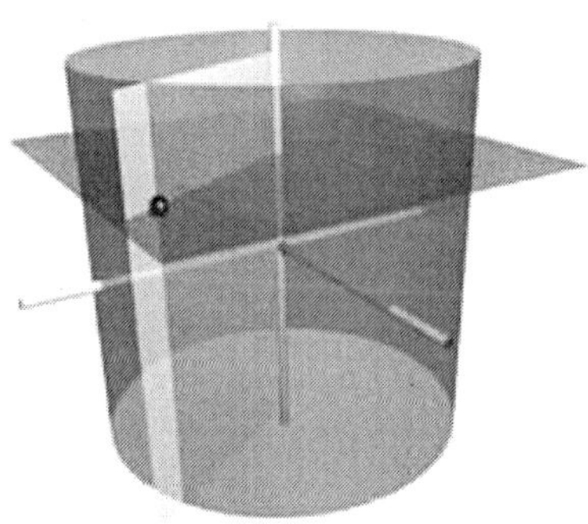

Figure: *The coordinate surfaces of the cylindrical coordinates (ρ, φ, z). The red cylinder shows the points with ρ=2, the blue plane shows the points with z=1, and the yellow half-plane shows the points with φ="60°. The z-axis is vertical and the x-axis is highlighted in green. The three surfaces intersect at the point P with those coordinates (shown as a black sphere); the Cartesian coordinates of P are roughly (1.0, "1.732, 1.0).*

The notation for cylindrical coordinates is not uniform. The ISO standard 31-11 recommends (ρ, φ, z), where ρ is the radial coordinate, φ the azimuth, and z the height. However, the radius is also often denoted r, the azimuth by θ or t, and the third coordinate by h or (if the cylindrical axis is considered horizontal) x, or any context-specific letter.

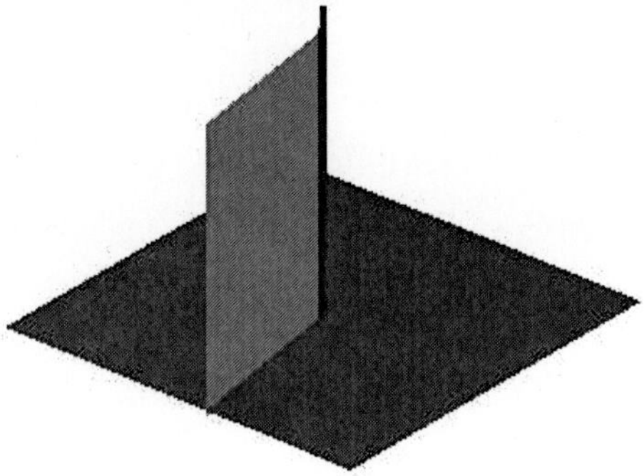

Figure: *Cylindrical Coordinate Surfaces. The three orthogonal components, ρ (green), φ (red), and z (blue), each increasing at a constant rate. The point is at the intersection between the three coloured surfaces.*

In concrete situations, and in many mathematical illustrations, a positive angular coordinate is measured counterclockwise as seen from any point with positive height.

Coordinate System Conversions

The cylindrical coordinate system is one of many three-dimensional coordinate systems. The following formula may be used to convert between them.

Cartesian Coordinates

For the conversion between cylindrical and Cartesian coordinate systems, it is convenient to assume that the reference plane of the former is the Cartesian x–y plane (with equation $z = 0$), and the cylindrical axis is the Cartesian z axis. Then the z coordinate is the same in both systems, and the correspondence between cylindrical (ρ,φ) and Cartesian (x,y) are the same as for polar coordinates, namely

$$x = \rho\cos\varphi$$

$y = \rho\sin\varphi$ in one direction, and

$$\rho = \sqrt{x^2 + y^2}$$

$$\varphi = \begin{cases} 0 & \text{if } x = 0 \text{ and } y = 0 \\ \arcsin(\frac{y}{\rho}) & \text{if } x \geq 0 \\ -\arcsin(\frac{y}{\rho}) + \pi & \text{if } x < 0 \end{cases}$$

in the other. The arcsin function is the inverse of the sine function, and is assumed to return an angle in the range [-π/2,+π/2] = [-90°,+90°]. These formulas yield an azimuth φ in the range [“90°,+270°].

Many modern programming languages provide a function that will compute the correct azimuth φ, in the range (-π, π], given x and y, without the need to perform a case analysis as above. For example, this function is called by atan2(y,x) in the C programming language, and atan(y,x) in Common Lisp.

Chapter 11

Basic Concepts and Operations

In vector algebra the basic objects are scalars and vectors, and the operations (beyond the vector space operations of scalar multiplication and vector addition) are the dot (or scalar) product and the cross product $\times$.

In geometric algebra the basic objects are multivectors (scalars are 0-vectors, vectors are 1-vectors, etc.), and the operations include the Clifford product (here called "geometric product") and the exterior product. The dot product/inner product/scalar product is defined on 1-vectors, and allows the geometric product to be expressed as the sum of the inner product and the exterior product when multiplying 1-vectors.

A distinguishing feature is that vector algebra uses the cross product, while geometric algebra uses the exterior product (and the geometric product). More subtly, geometric algebra in Euclidean 3-space distinguishes 0-vectors, 1-vectors, 2-vectors, and 3-vectors, while elementary vector algebra identifies 1-vectors and 2-vectors (as vectors) and 0-vectors and 3-vectors (as scalars), though more advanced vector algebra distinguishes these as scalars, vectors, pseudovectors, and pseudoscalars. Unlike vector algebra, geometric algebra includes sums of k-vectors of differing k.

The cross product does not generalize to dimensions other than 3 (as a product of two vectors, yielding a third vector), and in higher dimensions not all k-vectors can be identified with vectors or scalars. By contrast, the exterior product (and geometric product) is defined uniformly for all dimensions and signatures, and multivectors are closed under these operations.

Embellishments, ad Hoc Techniques, and Tricks

More advanced treatments of vector algebra add embellishments to the initial picture – pseudovectors and pseudoscalars (in geometric algebra terms, 2-vectors and 3-vectors), while applications to other dimensions use ad hoc techniques and "tricks" rather than a general mathematical approach. By contrast, geometric algebra begins with a complete picture, and applies uniformly in all dimensions.

For example, applying vector calculus in 2 dimensions, such as to compute torque or curl, requires adding an artificial 3rd dimension and extending the vector field to be constant in that dimension.

The torque or curl is then a normal vector field in this 3rd dimension. By contrast, geometric algebra in 2 dimensions defines torque and curl as pseudoscalar fields (2-vector fields), without requiring a 3rd dimension. Similarly, the scalar triple product is ad hoc, and can instead be expressed uniformly using the exterior product and the geometric product.

List of Analogous Formulas

Here are some comparisons between standard $\mathbb{R}^3$ vector relations and their corresponding wedge product and geometric product equivalents. All the wedge and geometric product equivalents here are good for more than three dimensions, and some also for two.

In two dimensions the cross product is undefined even if what it describes (like torque) is perfectly well defined in a plane without introducing an arbitrary normal vector outside of the space. Many of these relationships only require the introduction of the wedge product to generalize, but since that may not be familiar to somebody with only a traditional background in vector algebra and calculus, some examples are given.

Algebraic and Geometric Properties of Cross and Wedge Products

Cross and wedge products are both antisymmetric:

$$v \times u = -(u \times v)$$

$$v \wedge u = -(u \wedge v)$$

They are both linear in the first operand

$$(u+v)\times w = u\times w + v\times w$$

$$(u+v)\wedge w = u\wedge w + v\wedge w$$

and in the second operand

$$u\times(v+w) = u\times v + u\times w$$

$$u\wedge(v+w) = u\wedge v + u\wedge w$$

In general, the cross product is not associative, while the wedge product is

$$(u\times v)\times w \neq u\times(v\times w)$$

$$(u\wedge v)\wedge w = u\wedge(v\wedge w)$$

Both the cross and wedge products of two identical vectors are zero:

$$u\times u = 0$$

$$u\wedge u = 0$$

$u\times v$ is perpendicular to the plane containing u and v.

$u\wedge v$ is an oriented representation of the same plane.

The cross product of traditional vector algebra (on $\mathbb{R}^3$) find its place in geometric algebra $\mathcal{G}_3$ as a scaled exterior product

$$a\times b = -i(a\wedge b)$$

(this is antisymmetric). Relevant is the distinction between axial and polar vectors in vector algebra, which is natural in geometric algebra as the distinction between vectors and bivectors (elements of grade two).

The here is a unit pseudoscalar of Euclidean 3-space, which establishes a duality between the vectors and the bivectors, and is named so because of the expected property

$$i^2 = (e_1e_2e_3)^2 = e_1e_2e_3e_1e_2e_3 = -e_1e_2e_1e_3e_2e_3 = e_1e_1e_2e_3e_2e_3 = -e_3e_2e_2e_3 = -1$$

The equivalence of the $\mathbb{R}^3$ cross product and the wedge product expression above can be confirmed by direct multiplication of $-i = -e_1e_2e_3$ with a determinant expansion of the wedge product

$$u\wedge v = \sum_{1\le i<j\le 3}(u_iv_j - v_iu_j)e_i\wedge e_j = \sum_{1\le i<j\le 3}(u_iv_j - v_iu_j)e_ie_j$$

Essentially, the geometric product of a bivector and the pseudoscalar of Euclidean 3-space provides a method of calculation of the Hodge dual.

Norm of a Vector

The norm (length) of a vector is defined in terms of the dot product

$$||u||^2 = u \cdot u$$

Using the geometric product this is also true, but this can be also be expressed more compactly as

$$||u||^2 = u^2$$

This follows from the definition of the geometric product and the fact that a vector wedge product with itself is zero

$$uu = u \cdot u + u \wedge u = u \cdot u$$

Lagrange Identity

In three dimensions the product of two vector lengths can be expressed in terms of the dot and cross products

$$||u||^2||v||^2 = (u \cdot v)^2 + ||u \times v||^2$$

The corresponding generalization expressed using the geometric product is

$$||u||^2||v||^2 = (u \cdot v)^2 - (u \wedge v)^2$$

This follows from expanding the geometric product of a pair of vectors with its reverse

$$(uv)(vu) = (u \cdot v + u \wedge v)(u \cdot v - u \wedge v)$$

Determinant Expansion of Cross and Wedge Products

$$u \times v = \sum_{i<j} \begin{vmatrix} u_i & u_j \\ v_i & v_j \end{vmatrix} e_i \times e_j$$

$$u \wedge v = \sum_{i<j} \begin{vmatrix} u_i & u_j \\ v_i & v_j \end{vmatrix} e_i \wedge e_j$$

Without justification or historical context, traditional linear algebra texts will often define the determinant as the first step of an elaborate sequence of definitions and theorems leading up to the solution of linear systems, Cramer's rule and matrix inversion.

An alternative treatment is to axiomatically introduce the wedge product, and then demonstrate that this can be used directly to solve linear systems. This is shown below, and does not require sophisticated math skills to understand.

It is then possible to define determinants as nothing more than the coefficients of the wedge product in terms of "unit k-vectors" ($e_i \wedge e_j$ terms) expansions as above.

- A one by one determinant is the coefficient of e_1 for an $\mathbb{R}^1$ 1-vector.
- A two-by-two determinant is the coefficient of $e_1 \wedge e_2$ for an $\mathbb{R}^2$ bivector
- A three-by-three determinant is the coefficient of $e_1 \wedge e_2 \wedge e_3$ for an $\mathbb{R}^3$ trivector...

When linear system solution is introduced via the wedge product, Cramer's rule follows as a side effect, and there is no need to lead up to the end results with definitions of minors, matrices, matrix invertibility, adjoints, cofactors, Laplace expansions, theorems on determinant multiplication and row column exchanges, and so forth.

Matrix Related

Matrix inversion (Cramer's rule) and determinants can be naturally expressed in terms of the wedge product. The use of the wedge product in the solution of linear equations can be quite useful for various geometric product calculations.

Traditionally, instead of using the wedge product, Cramer's rule is usually presented as a generic algorithm that can be used to solve linear equations of the form $Ax = b$ (or equivalently to invert a matrix). Namely

$$x = \frac{1}{|A|}\mathrm{adj}(A)b\,.$$

This is a useful theoretic result. For numerical problems row reduction with pivots and other methods are more stable and efficient. When the wedge product is coupled with the Clifford product and put into a natural geometric context, the fact that the determinants are used in the expression of

$\mathbb{R}^N$ parallelogram area and parallelepiped volumes (and higher dimensional generalizations of these) also comes as a nice side effect. As is also shown below, results such as Cramer's rule also follow directly from the property of the wedge product that it selects non identical elements. The end result is then simple enough that it could be derived easily if required instead of having to remember or look up a rule.

Two variables example

$$\begin{bmatrix} a & b \end{bmatrix}\begin{bmatrix} x \\ y \end{bmatrix} = ax + by = c$$

Pre and post multiplying by a and b

$$(ax + by) \wedge b = (a \wedge b)x = c \wedge b$$

$$a \wedge (ax + by) = (a \wedge b)y = a \wedge c$$

Provided $a \wedge b \neq 0$ the solution is

$$\begin{bmatrix} x \\ y \end{bmatrix} = \frac{1}{a \wedge b}\begin{bmatrix} c \wedge b \\ a \wedge c \end{bmatrix}$$

For $a, b \in \mathbb{R}^2$, this is Cramer's rule since the $e_1 \wedge e_2$ factors of the wedge products

$$u \wedge v = \begin{vmatrix} u_1 & u_2 \\ v_1 & v_2 \end{vmatrix} e_1 \wedge e_2$$

divide out.

Similarly, for three, or N variables, the same ideas hold

$$\begin{bmatrix} a & b & c \end{bmatrix}\begin{bmatrix} x \\ y \\ z \end{bmatrix} = d$$

$$\begin{bmatrix} x \\ y \\ z \end{bmatrix} = \frac{1}{a \wedge b \wedge c}\begin{bmatrix} d \wedge b \wedge c \\ a \wedge d \wedge c \\ a \wedge b \wedge d \end{bmatrix}$$

Again, for the three variable three equation case this is Cramer's rule since the $e_1 \wedge e_2 \wedge e_3$ factors of all the wedge products divide out, leaving the familiar determinants.

A numeric example with three equations and two unknowns When there are more equations than variables case, if the equations have a solution, each of the k-vector quotients will be scalars

To illustrate here is the solution of a simple example with three equations and two unknowns.

$$\begin{bmatrix}1\\1\\0\end{bmatrix}x+\begin{bmatrix}1\\1\\1\end{bmatrix}y=\begin{bmatrix}1\\1\\2\end{bmatrix}$$

The right wedge product with $(1,1,1)$ solves for x

$$\begin{bmatrix}1\\1\\0\end{bmatrix}\wedge\begin{bmatrix}1\\1\\1\end{bmatrix}x=\begin{bmatrix}1\\1\\2\end{bmatrix}\wedge\begin{bmatrix}1\\1\\1\end{bmatrix}$$

and a left wedge product with $(1,1,0)$ solves for y

$$\begin{bmatrix}1\\1\\0\end{bmatrix}\wedge\begin{bmatrix}1\\1\\1\end{bmatrix}y=\begin{bmatrix}1\\1\\0\end{bmatrix}\wedge\begin{bmatrix}1\\1\\2\end{bmatrix}.$$

Observe that both of these equations have the same factor, so one can compute this only once (if this was zero it would indicate the system of equations has no solution).

Collection of results for and yields a Cramers rule like form:

$$\begin{bmatrix}x\\y\end{bmatrix}=\frac{1}{(1,1,0)\wedge(1,1,1)}\begin{bmatrix}(1,1,2)\wedge(1,1,1)\\(1,1,0)\wedge(1,1,2)\end{bmatrix}.$$

Writing $e_i\wedge e_j=e_{ij}$, we have the end result:

$$\begin{bmatrix}x\\y\end{bmatrix}=\frac{1}{e_{13}+e_{23}}\begin{bmatrix}-e_{13}-e_{23}\\2e_{13}+2e_{23}\end{bmatrix}=\begin{bmatrix}-1\\2\end{bmatrix}.$$

Equation of a Plane

For the plane of all points r through the plane passing through three independent points r_1, r_0, and r_2, the normal form of the equation is

$$((r_2 - r_0) \times (r_1 - r_0)) \cdot (r - r_0) = 0.$$

The equivalent wedge product equation is

$$(r_2 - r_0) \wedge (r_1 - r_0) \wedge (r - r_0) = 0.$$

Projection and Rejection

Using the Gram–Schmidt process a single vector can be decomposed into two components with respect to a reference vector, namely the projection onto a unit vector in a reference direction, and the difference between the vector and that projection.

With, $\hat{u} = u / \|u\|$, the projection of onto is

$$\mathrm{Proj}_{\hat{u}}\, v = \hat{u}(\hat{u} \cdot v)$$

Orthogonal to that vector is the difference, designated the rejection,

$$v - \hat{u}(\hat{u} \cdot v) = \frac{1}{\|u\|^2}(\|u\|^2\, v - u(u \cdot v))$$

The rejection can be expressed as a single geometric algebraic product in a few different ways

$$\frac{u}{u^2}(uv - u \cdot v) = \frac{1}{u}(u \wedge v) = \hat{u}(\hat{u} \wedge v) = (v \wedge \hat{u})\hat{u}$$

The similarity in form between the projection and the rejection is notable. The sum of these recovers the original vector

$$v = \hat{u}(\hat{u} \cdot v) + \hat{u}(\hat{u} \wedge v)$$

Here the projection is in its customary vector form. An alternate formulation is possible that puts the projection in a form that differs from the usual vector formulation

$$v = \frac{1}{u}(u \cdot v) + \frac{1}{u}(u \wedge v) = (v \cdot u)\frac{1}{u} + (v \wedge u)\frac{1}{u}$$

Working backwards from the end result, it can be observed that this orthogonal decomposition result can in fact follow more directly from the definition of the geometric product itself.

$$v = \hat{u}\hat{u}v = \hat{u}(\hat{u} \cdot v + \hat{u} \wedge v)$$

With this approach, the original geometrical consideration is not necessarily obvious, but it is a much quicker way to get at the same algebraic result.

However, the hint that one can work backwards, coupled with the knowledge that the wedge product can be used to solve sets of linear equations, the problem of orthogonal decomposition can be posed directly,

Let $v = au + x$, where $u \cdot x = 0$. To discard the portions of that are collinear with , take the wedge product

$$u \wedge v = u \wedge (au + x) = u \wedge x$$

Here the geometric product can be employed

$$u \wedge v = u \wedge x = ux - u \cdot x = ux$$

Because the geometric product is invertible, this can be solved for x

$$x = \frac{1}{u}(u \wedge v)$$

The same techniques can be applied to similar problems, such as calculation of the component of a vector in a plane and perpendicular to the plane.

For three dimensions the projective and rejective components of a vector with respect to an arbitrary non-zero unit vector, can be expressed in terms of the dot and cross product

$$v = (v \cdot \hat{u})\hat{u} + \hat{u} \times (v \times \hat{u})$$

For the general case the same result can be written in terms of the dot and wedge product and the geometric product of that and the unit vector

$$v = (v \cdot \hat{u})\hat{u} + (v \wedge \hat{u})\hat{u}$$

It's also worthwhile to point out that this result can also be expressed using right or left vector division as defined by the geometric product

$$v = (v \cdot u)\frac{1}{u} + (v \wedge u)\frac{1}{u}$$

$$v = \frac{1}{u}(u \cdot v) + \frac{1}{u}(u \wedge v)$$

Like vector projection and rejection, higher dimensional analogs of that calculation are also possible using the geometric product.

As an example, one can calculate the component of a vector perpendicular to a plane and the projection of that vector onto the plane. Let $w = au + bv + x$, where $u \cdot x = v \cdot x = 0$. As above, to discard the portions of that are colinear with v or v, take the wedge product

$$w \wedge u \wedge v = (au + bv + x) \wedge u \wedge v = x \wedge u \wedge v.$$

Having done this calculation with a vector projection, one can guess that this quantity equals $x(u \wedge v)$. One can also guess there is a vector and bivector dot product like quantity such that the allows the calculation of the component of a vector that is in the "direction of a plane".

Both of these guesses are correct, and validating these facts is worthwhile. However, skipping ahead slightly, this to-be-proved fact allows for a nice closed form solution of the vector component outside of the plane:

$$x = (w \wedge u \wedge v)\frac{1}{u \wedge v} = \frac{1}{u \wedge v}(u \wedge v \wedge w).$$

Notice the similarities between this planar rejection result a the vector rejection result. To calculation the component of a vector outside of a plane we take the volume spanned by three vectors (trivector) and "divide out" the plane. Independent of any use of the geometric product it can be shown that this rejection in terms of the standard basis is

$$x = \frac{1}{(A_{u,v})^2} \sum_{i<j<k} \begin{vmatrix} w_i & w_j & w_k \\ u_i & u_j & u_k \\ v_i & v_j & v_k \end{vmatrix} \begin{vmatrix} u_i & u_j & u_k \\ v_i & v_j & v_k \\ e_i & e_j & e_k \end{vmatrix}$$

where

$$(A_{u,v})^2 = \sum_{i<j} \begin{vmatrix} u_i & u_j \\ v_i & v_j \end{vmatrix} = -(u \wedge v)^2$$

is the squared area of the parallelogram formed by , and .

The (squared) magnitude of x is

$$\|x\|^2 = x \cdot w = \frac{1}{(A_{u,v})^2} \sum_{i<j<k} \begin{vmatrix} w_i & w_j & w_k \\ u_i & u_j & u_k \\ v_i & v_j & v_k \end{vmatrix}^2$$

Thus, the (squared) volume of the parallelopiped (base area times perpendicular height) is

$$\sum_{i<j<k} \begin{vmatrix} w_i & w_j & w_k \\ u_i & u_j & u_k \\ v_i & v_j & v_k \end{vmatrix}^2$$

Note the similarity in form to the w, u, v trivector itself

$$\sum_{i<j<k} \begin{vmatrix} w_i & w_j & w_k \\ u_i & u_j & u_k \\ v_i & v_j & v_k \end{vmatrix} e_i \wedge e_j \wedge e_k$$

which, if you take the set of $e_i \wedge e_j \wedge e_k$ as a basis for the trivector space, suggests this is the natural way to define the length of a trivector. Loosely speaking the length of a vector is a length, length of a bivector is area, and the length of a trivector is volume.

If a vector is factored directly into projective and rejective terms using the geometric product $v = \frac{1}{u}(u \cdot v + u \wedge v)$, then it is not necessarily obvious that the rejection term, a product of vector and bivector is even a vector. Expansion of the vector bivector product in terms of the standard basis vectors has the following form

Let $r = \frac{1}{u}(u \wedge v) = \frac{u}{u^2}(u \wedge v) = \frac{1}{\|u\|^2} u(u \wedge v)$

It can be shown that

$$r = \frac{1}{\|u\|^2} \sum_{i<j} \begin{vmatrix} u_i & u_j \\ v_i & v_j \end{vmatrix} \begin{vmatrix} u_i & u_j \\ e_i & e_j \end{vmatrix}$$

(a result that can be shown more easily straight from $r = v - \hat{u}(\hat{u} \cdot v)$).

The rejective term is perpendicular to u, since $\begin{vmatrix} u_i & u_j \\ u_i & u_j \end{vmatrix} = 0$ implies $r \cdot u = 0$.

The magnitude of r, is

$$||r||^2 = r\cdot v = \frac{1}{||u||^2}\sum_{i<j}\begin{vmatrix} u_i & u_j \\ v_i & v_j \end{vmatrix}^2 .$$

So, the quantity

$$||r||^2||u||^2 = \sum_{i<j}\begin{vmatrix} u_i & u_j \\ v_i & v_j \end{vmatrix}^2$$

is the squared area of the parallelogram formed by and.

It is also noteworthy that the bivector can be expressed as

$$u\wedge v = \sum_{i<j}\begin{vmatrix} u_i & u_j \\ v_i & v_j \end{vmatrix} e_i \wedge e_j .$$

Thus is it natural, if one considers each term $e_i \wedge e_j$ as a basis vector of the bivector space, to define the (squared) "length" of that bivector as the (squared) area.

Going back to the geometric product expression for the length of the rejection $\frac{1}{u}(u\wedge v)$ we see that the length of the quotient, a vector, is in this case is the "length" of the bivector divided by the length of the divisor.

This may not be a general result for the length of the product of two k-vectors, however it is a result that may help build some intuition about the significance of the algebraic operations. Namely,

When a vector is divided out of the plane (parallelogram span) formed from it and another vector, what remains is the perpendicular component of the remaining vector, and its length is the planar area divided by the length of the vector that was divided out.

Area of the Parallelogram Defined by u and V

If A is the area of the parallelogram defined by u and v, then

$$A^2 = ||u\times v||^2 = \sum_{i<j}\begin{vmatrix} u_i & u_j \\ v_i & v_j \end{vmatrix}^2 ,$$

and

$$A^2 = -(u \wedge v)^2 = \sum_{i<j} \begin{vmatrix} u_i & u_j \\ v_i & v_j \end{vmatrix}^2.$$

Note that this squared bivector is a geometric multiplication; this computation can alternatively be stated as the Gram determinant of the two vectors.

Angle Between two Vectors

$$(\sin\theta)^2 = \frac{||u \times v||^2}{||u||^2||v||^2}$$

$$(\sin\theta)^2 = -\frac{(u \wedge v)^2}{u^2 v^2}$$

Volume of the Parallelopiped Formed by Three Vectors

In vector algebra, the volume of a parallelopiped is given by the square root of the squared norm of the scalar triple product:

$$V^2 = ||(u \times v) \cdot w||^2 = \begin{vmatrix} u_1 & u_2 & u_3 \\ v_1 & v_2 & v_3 \\ w_1 & w_2 & w_3 \end{vmatrix}^2$$

$$V^2 = -(u \wedge v \wedge w)^2 = -\left(\sum_{i<j<k} \begin{vmatrix} u_i & u_j & u_k \\ v_i & v_j & v_k \\ w_i & w_j & w_k \end{vmatrix} \hat{e}_i \wedge \hat{e}_j \wedge \hat{e}_k \right)^2$$

$$= \sum_{i<j<k} \begin{vmatrix} u_i & u_j & u_k \\ v_i & v_j & v_k \\ w_i & w_j & w_k \end{vmatrix}^2$$

Product of a Vector and a Bivector

In order to justify the normal to a plane result above, a general examination of the product of a vector and bivector is required. Namely,

$$w(u \wedge v) = \sum_{i,j<k} w_i e_i \begin{vmatrix} u_j & u_k \\ v_j & v_k \end{vmatrix} e_j \wedge e_k$$

This has two parts, the vector part where $i = j$ or $i = k$, and the trivector parts where no indexes equal. After some index summation trickery, and grouping terms and so forth, this is

$$w(u \wedge v) = \sum_{i<j} (w_i e_j - w_j e_i) \begin{vmatrix} u_i & u_j \\ v_i & v_j \end{vmatrix} + \sum_{i<j<k} \begin{vmatrix} w_i & w_j & w_k \\ u_i & u_j & u_k \\ v_i & v_j & v_k \end{vmatrix} e_i \wedge e_j \wedge e_k$$

The trivector term is $w \wedge u \wedge v$. Expansion of $(u \wedge v)w$ yields the same trivector term (it is the completely symmetric part), and the vector term is negated. Like the geometric product of two vectors, this geometric product can be grouped into symmetric and antisymmetric parts, one of which is a pure k-vector.

In analogy the antisymmetric part of this product can be called a generalized dot product, and is roughly speaking the dot product of a "plane" (bivector), and a vector.

The properties of this generalized dot product remain to be explored, but first here is a summary of the notation

$$w(u \wedge v) = w \cdot (u \wedge v) + w \wedge u \wedge v$$

$$(u \wedge v)w = -w \cdot (u \wedge v) + w \wedge u \wedge v$$

$$w \wedge u \wedge v = \frac{1}{2}(w(u \wedge v) + (u \wedge v)w)$$

$$w \cdot (u \wedge v) = \frac{1}{2}(w(u \wedge v) - (u \wedge v)w)$$

Let $w = x + y$, where $x = au + bv$, and $y \cdot u = y \cdot v = 0$. Expressing w and the $u \wedge v$, products in terms of these components is

$$w(u \wedge v) = x(u \wedge v) + y(u \wedge v) = x \cdot (u \wedge v) + y \cdot (u \wedge v) + y \wedge u \wedge v$$

With the conditions and definitions above, and some manipulation, it can be shown that the term $y \cdot (u \wedge v) = 0$, which then justifies the previous solution of the normal to a plane problem.

Since the vector term of the vector bivector product the name dot product is zero when the vector is perpendicular to the plane (bivector), and this vector, bivector "dot product" selects only the components that are in the plane, so in analogy to the vector-vector dot product this name itself is justified by more than the fact this is the non-wedge product term of the geometric vector-bivector product.

Derivative of a Unit Vector

It can be shown that a unit vector derivative can be expressed using the cross product

$$\frac{d}{dt}\left(\frac{r}{||r||}\right)=\frac{1}{||r||^3}\left(r\times\frac{dr}{dt}\right)\times r=\left(\hat{r}\times\frac{1}{||r||}\frac{dr}{dt}\right)\times\hat{r}$$

The equivalent geometric product generalization is

$$\frac{d}{dt}\left(\frac{r}{||r||}\right)=\frac{1}{||r||^3}r\left(r\wedge\frac{dr}{dt}\right)=\frac{1}{r}\left(\hat{r}\wedge\frac{dr}{dt}\right)$$

Thus this derivative is the component of $\frac{1}{||r||}\frac{dr}{dt}$ in the direction perpendicular to r. In other words this is $\frac{1}{||r||}\frac{dr}{dt}$ minus the projection of that vector onto $\hat{r}$.

This intuitively makes sense (but a picture would help) since a unit vector is constrained to circular motion, and any change to a unit vector due to a change in its generating vector has to be in the direction of the rejection of $\hat{r}$ from $\frac{dr}{dt}$. That rejection has to be scaled by 1/|r| to get the final result.

When the objective isn't comparing to the cross product, it's also notable that this unit vector derivative can be written

$$r\frac{d\hat{r}}{dt}=\hat{r}\wedge\frac{dr}{dt}$$

Complex Lamellar Vector Field

In vector calculus, a complex lamellar vector field is a vector field in three dimensions which is orthogonal to its own curl. That is,

$$F\cdot(\nabla\times F)=0.$$

Complex lamellar vector fields are precisely those that are normal to a family of surfaces. A special case are irrotational vector fields, satisfying $\nabla\times F=0.$

An irrotational vector field is locally the gradient of a function, and is therefore orthogonal to the family of level surfaces (the equipotential surfaces). Accordingly, the term lamellar vector field is sometimes used as a synonym for an irrotational vector field. The adjective "lamellar" derives from the noun "lamella", which means a thin layer. The *lamellae* to which "lamellar flow" refers are the surfaces of constant potential, or in the complex case, the surfaces orthogonal to the vector field.

Concatenation (Mathematics)

In mathematics, concatenation is the joining of two numbers by their numerals. That is, the concatenation of 123 and 456 is 123456. Concatenation of numbers a and b is denoted $a \mid\mid b$. Relevant subjects in recreational mathematics include Smarandache-Wellin numbers, home primes, and Champernowne's constant.

The convention for sequences at places such as the Online Encyclopedia of Integer Sequences is to have sequences of concatenations include as the first term a number prior to the actual act of concatenation.

Therefore, care must be taken to ensure that parties discussing a topic agree either with this convention or with plain language. For example, the first term in the sequence of concatenations of increasing even numbers may be taken to be either 24, as would seem obviously correct, or simply 2, according to convention.

Calculation

The concatenation of numbers depends on the numeric base, which is often understood from context.

Given the numbers p and q in base b, the concatenation $p \mid\mid q$ is given by

$$p \mid\mid q = pb^{l(q)} + q$$

where $l(q) = \mid\mid log_b(q) \mid\mid + 1$

is the number of digits of q in base b, and $[x]$ is the floor function.

Vector Extension

The concatenation of vectors can be understood in two distinct ways; either as a generalization of the above operation for numbers or as a concatenation of lists.

Given two vectors in $\mathbb{R}^n$, concatenation can be defined as

$$\begin{pmatrix} a_1 \\ a_2 \\ \vdots \\ a_n \end{pmatrix} || \begin{pmatrix} b_1 \\ b_2 \\ \vdots \\ b_n \end{pmatrix} = \begin{pmatrix} a_1 || b_1 \\ a_2 || b_2 \\ \vdots \\ a_n || b_n \end{pmatrix}$$

In the case of vectors in $\mathbb{R}^1$, this is equivalent to the above definition for numbers. The further extension to matrices is trivial. Since vectors can be viewed in a certain way as lists, concatenation may take on another meaning. In this case the concatenation of two lists $(a_1, a_2, ..., a_n)$ and $(b_1, b_2, ..., b_n)$ is the list $(a_1, a_2, ..., a_n, b_1, b_2, ..., b_n)$. Only the exact context will reveal which meaning is intended.

d'Alembert–Euler Condition

In mathematics and physics, especially the study of mechanics and fluid dynamics, the d'Alembert-Euler condition is a requirement that the streaklines of a flow are irrotational. Let x = x(X,t) be the coordinates of the point x into which X is carried at time t by a (fluid) flow. Let $\ddot{x} = \frac{D^2x}{Dt}$ be the second material derivative of x. Then the d'Alembert-Euler condition is:

$$\text{curl } x = 0.$$

The d'Alembert-Euler condition is named for Jean le Rond d'Alembert and Leonhard Euler who independently first described its use in the mid-18th century. It is not to be confused with the Cauchy-Riemann conditions.

Del

In vector calculus, del is a vector differential operator, usually represented by the nabla symbol ". When applied to a function defined on a one-dimensional domain, it denotes its standard derivative as defined in calculus. When applied to a field (a function defined on a multidimensional domain), del may denote the gradient (locally steepest slope) of a scalar field, the divergence of a vector field, or the curl (rotation) of a vector field, depending on the way it is applied. Strictly speaking, del is not a specific operator, but rather a convenient mathematical notation for those three operators, that makes many equations easier to write and remember.

The del symbol can be interpreted as a vector of partial derivative operators, and its three possible meanings—gradient, divergence, and curl—can be formally viewed as the product of scalars, dot product, and cross product, respectively, of the del "operator" with the field. These formal products do not necessarily commute with other operators or products.

Definition

In the three-dimensional Cartesian coordinate system R^3 with coordinates (x, y, z), del is defined in terms of partial derivative operators as

$$\nabla = \hat{x}\frac{\partial}{\partial x} + \hat{y}\frac{\partial}{\partial y} + \hat{z}\frac{\partial}{\partial z}$$

where $\{\hat{x}, \hat{y}, \hat{z}\}$ are the unit vectors in their respective directions. Though this page chiefly treats del in three dimensions, this definition can be generalized to the n-dimensional Euclidean space R^n. In the Cartesian coordinate system with coordinates $(x_1, x_2, ..., x_n)$, del is:

$$\nabla = \sum_{i=1}^{n} \hat{e}^i \frac{\partial}{\partial x_i}$$

where $\{\hat{e}^i : 1 \leq i \leq n\}$ is the standard basis in this space.

More compactly, using the Einstein summation notation, del is written as

$$\nabla = \hat{e}^i \partial_i$$

Notational Uses

Del is used as a shorthand form to simplify many long mathematical expressions. It is most commonly used to simplify expressions for the gradient, divergence, curl, directional derivative, and Laplacian.

Gradient

The vector derivative of a scalar field f is called the gradient, and it can be represented as:

$$\nabla f = \frac{\partial f}{\partial x}\hat{x} + \frac{\partial f}{\partial y}\hat{y} + \frac{\partial f}{\partial z}\hat{z}$$

It always points in the direction of greatest increase of f, and it has a magnitude equal to the maximum rate of increase at the point—just like a standard derivative.

In particular, if a hill is defined as a height function over a plane $h(x,y)$, the 2d projection of the gradient at a given location will be a vector in the xy-plane (sort of like an arrow on a map) pointing along the steepest direction. The magnitude of the gradient is the value of this steepest slope.

In particular, this notation is powerful because the gradient product rule looks very similar to the 1d-derivative case:

$$\nabla(fg) = f\nabla g + g\nabla f$$

However, the rules for dot products do not turn out to be simple, as illustrated by:

$$\nabla(\vec{u}\cdot\vec{v}) = (\vec{u}\cdot\nabla)\vec{v} + (\vec{v}\cdot\nabla)\vec{u} + \vec{u}\times(\nabla\times\vec{v}) + \vec{v}\times(\nabla\times\vec{u}) .$$

Bibliography

Boyd, John P., *Chebyshev and Fourier Spectral Methods,* Dover, 2019.

Carslaw, H.S.; and Jaeger, J.C., *Conduction of Heat in Solids*, Oxford University Press, 1999.

Chris Doran and Anthony Lasenby, *Geometric Algebra for Physicists,* Cambridge University Press, Cambridge, 2013.

Crowe, M.J., *A History of Vector Analysis*, Notre Dame University Press, 2007.

Dautray R. and Lions, J. L., *Spectral Theory and Applications,* Springer-Verlag, 2010.

Dennis Zill, G.; Michael Cullen R., *Advanced engineering mathematics,* Jones & Bartlett Learning, 2016.

Dummit, David S.; Foote, Richard M., *Abstract Algebra*, John Wiley & Sons, 2014.

E. A. Milne, *Vectorial Mechanics*, Methuen Publishing, London, 1998.

George Arfken B. and Hans Weber, J., *Mathematical Methods for Physicists*, Academic Press, 2015.

Girault V. and Raviart, P.A., *Finite Element Methods for Navier–Stokes Equations: Theory and Algorithms,* Springer-Verlag, 2006.

Jacques Willy Delleur, *The handbook of groundwater engineering,* CRC Press, 2016.

Jeffreys, H. and Jeffreys, BS., *Methods of mathematical physics,* Cambridge University Press, 2019.

John Hopcroft E. and Jeffrey Ullman, D., *Introduction to Automata Theory, Languages and Computation,* Addison-Wesley Publishing, Reading Massachusetts, 2008.

Joseph, Pedlosky, *Geophysical fluid dynamics,* Springer, 2007.

Lass, Harry, *Vector and Tensor Analysis*, McGraw-Hill Book Company, 1990.

Lay, David C., *Linear Algebra and Its Applications*, Wesley, 2016.

Lounesto, Pertti, *Clifford algebras and spinors*, Cambridge University Press, Cambridge, 2019.

Marsden, J.E., *Vector Calculus*, W. H. Freeman & Company, 2006.

Maxwell, James Clerk, *Treatise on Electricity and Magnetism*, 1992.

Michael Crowe J., *A History of Vector Analysis : The Evolution of the Idea of a Vectorial System*, Dover Publications, 2007.

Nahin, Paul J., *Oliver Heaviside: the life, work, and times of an electrical genius of the Victorian age*, JHU Press, 2018.

Pertti Lounesto, *Clifford algebras and spinors*, Cambridge University Press, 2016.

Schey , H. M., *Div Grad Curl and all that: An informal text on vector calculus*, W. W. Norton & Company, 2015.

Serge, Lang, *Graduate Texts in Mathematics,* Springer, 2022.

Sheldon, Axler, *Linear Algebra Done Right,* Springer, 2022.

Strang, Gilbert, *Linear Algebra and Its Applications,* Brooks Cole, 2016.

Suthan Suthersan S., *Remediation engineering: design concepts*, CRC Press, 2006.

Weisstein, Eric W., *CRC concise encyclopedia of mathematics*, CRC Press, 2013.

Whelan, P.M., Hodgeson, M.J., *Essential Principles of Physics,* John Murray, 2018.

Index

E

G

H

P

Q

R

S

V

W

❑❑❑